BIOMECHANICS OF SPORT

A Research Approach

Doris I. Miller, Ph.D.

Associate Professor of Physical Education, University of Saskatchewan College of Physical Education, Saskatoon, Saskatchewan, Canada

Richard C. Nelson, Ph.D.

Professor of Physical Education, and Director, Biomechanics Laboratory, College of Health, Physical Education, and Recreation, The Pennsylvania State University University Park, Pennsylvania

Illustrated by ARNOLD J. GOLDFUSS

BIOMECHANICS OF SPORT

A Research Approach

LEA & FEBIGER · 1973 · PHILADELPHIA

Health Education, Physical Education,
and Recreation Series

RUTH ABERNATHY, Ph.D.
Editorial Adviser
Director, School of Physical and Health Education
University of Washington, Seattle

Library of Congress Cataloging in Publication Data

Miller, Doris I
Biomechanics of sport.

(Health education, physical education, and recreation series)
1. Sports—Physiological aspects. 2. Human mechanics. I. Nelson, Richard C., 1932- joint author. II. Title.
RC1235.M53 612'.76 73-3173
ISBN 0-8121-0431-5

ISBN 0-8121-0431-5

Library of Congress Catalog Number 73-3173

Published in Great Britain by Henry Kimpton Publishers, London

PRINTED IN THE UNITED STATES OF AMERICA

Preface

This text represents a departure from earlier books written on the topics of kinesiology and biomechanics and is indicative of the changes occurring in the subject matter and in the physical education curriculum. Kinesiology as traditionally taught at the undergraduate level has centered upon the anatomical and mechanical analysis of human movement. With the expansion of knowledge, however, it has become increasingly more difficult to cover both aspects adequately within a single course. As a result, there has been a trend toward the separation of these two areas of emphasis into different courses. This tendency has been paralleled by the emergence of new textbooks concerned primarily with the mechanical aspects of human motion.

Biomechanics of Sport: A Research Approach is a research-oriented publication intended primarily for senior undergraduate and graduate-level classes in physical education. Portions of it will also be suitable for introductory biomechanics courses. The text should meet the specific needs of graduate students in biomechanics and serve as a reference for researchers in the field. Persons in bioengineering, functional anatomy, industrial engineering, physical rehabilitation, ergonomics and related disciplines will find certain sections pertinent to their investigation of the mechanics of human performance.

This work represents an attempt to glean relevant material from a variety of content areas and to apply it to the biomechanics of sport. The selection of topics has been based upon the research and teaching experience of the authors combined with their careful review of existing literature. Because several texts which focus upon the mechanical components of specific sports skills are available or forthcoming, no attempt has been made to duplicate this material. Rather, the intention has been to fill an apparent need for a publication dealing with the content, research methodology and techniques, literature sources and related topics concerning the biomechanics of sport.

The authors wish to express their appreciation and gratitude to Andrew Pytel, Kenneth L. Petak and Mrs. Josephine Cleary for their valuable assistance in the review and preparation of the manuscript and to Arnold J. Goldfuss, Research Assistant, Biomechanics Laboratory, The Pennsylvania State University, for the illustrations.

Saskatoon, Saskatchewan, Canada DORIS I. MILLER

University Park, Pennsylvania RICHARD C. NELSON

Contents

CHAPTER 1

Introduction to The Biomechanics of Sport

BIOMECHANICS, the science which investigates the effects of internal and external forces upon living bodies, is by no means a new discipline. This fact is well documented in historical reviews published by Rasch (1958) and Contini and Drillis (1966a). What is relatively new, however, is the widespread interest in the application of biomechanics to the many facets of human motion. Contini and Drillis noted that following World War I research in ergonomics flourished in Germany and Russia. A marked increase in biomechanics inquiry in the United States immediately after World War II was initially stimulated by the transportation industry. Since the early 1950's, biomechanics has emerged as an important area of scientific investigation in a variety of disciplines. Included among these are space science, functional anatomy, automobile safety, orthopaedic surgery, biomedical engineering, physical rehabilitation, industrial psychology, aviation medicine, sport and physical education.

Until very recent years, progress in the biomechanics of sport was noticeably slow. Although Steindler made reference to the contribution which biomechanics could make to physical education in 1942 it was not until years later, in 1966, that Alley outlined the academic basis for this specialization and described the ingredients of a quality doctoral program. He proposed that the designation anthropomechanics replace the traditional label, kinesiology, which had become diluted in meaning and consequently had led to considerable confusion in the profession. During the 1960's other terms were also suggested, including homokinetics, kinanthropology, biokinetics and mechanical kinesiology. This problem in semantics appears now to have been resolved with the general acceptance of biomechanics as the descriptive label.

Evidence of the changing nature of sport biomechanics can be found in the expanding number of conferences, seminars and symposia being conducted. The First International Seminar on Biomechanics was held in Zurich, Switzerland in 1967. This meeting, sponsored by the Research Committee of the International Council of Sport and Physical Education of UNESCO, was a milestone for progress in the biomechanics of sport as it provided the first opportunity for interested persons from various parts of the world to meet and exchange ideas. The Second International Seminar, conducted in Eindhoven, Holland in 1969, was an outgrowth of the interest generated in Zurich. This biennial international seminar was continued in Rome in 1971 and appears to be well established, with the fourth meeting at The Pennsylvania State University in 1973. There have been other meetings which have focused on more specific topics. The Symposium on the Theory of Sport Technique was held in 1968 in Warsaw, Poland. Biomechanical factors influencing success in sport were emphasized during the three-day conference. A symposium on biomechanics in swimming, conducted in Brussels in 1970, was the first concerned with one area of sport. Shortly thereafter, the first meeting dealing with the biomechanics of sport in North America was held at Indiana University. This symposium was organized under the auspices of the Big Ten Committee on Institutional Cooperation in Physical Education. A smaller conference at The Pennsylvania State University in 1971 considered selected aspects of cinematography, course content, laboratory development and related topics. Short-term workshops have also been conducted on computer applications at Purdue University, electromyography at the University of Wisconsin and cinematography at The Pennsylvania State University.

The fact that the field of the biomechanics of sport is growing rapidly is also substantiated by the increase in literature, research laboratories and graduate programs. Several new books have been published which depart from the traditional major emphasis upon applied anatomy. These new texts, which stress the mechanical analysis of human motion, are the result of the expansion of knowledge in the field. A few books dealing with the biomechanics of specific sports such as diving, gymnastics skills and running have also appeared. These will proliferate in the future as research adds to our knowledge and teachers and coaches become better able to understand and implement their findings in practice. At the core of this new knowledge is the expanded research literature concerned with biomechanics. Journals in physical education and related fields are publishing an increasing number of papers concerned with this area of study.

Until the latter part of the 1960's, few opportunities were available in the United States for advanced graduate students to specialize in biomechanics. The content to support this work had not been clearly defined and few on-going research programs were in operation. This situation has changed rapidly so that now a number of doctoral programs and research laboratories have been developed. The contributions of young, well-qualified doctoral graduates now entering the field should result in an accelerated growth of biomechanics of sport.

Research in this area is characterized by two major thrusts. The first is the biological aspect in which functional anatomy is emphasized and electromyography is the principal research tool. Results of such investigations have enhanced the understanding of skeletal muscle activity during a variety of human movements. The formation of the International Society of Electromyographic Kinesiology (ISEK) and the increasing number of publications reflect the development of this sector of biomechanics.

The second thrust, which is the one stressed in this text, concerns the mechanical aspects of human motion. Research is conducted at three levels along a continuum; practical, fundamental and theoretical. The first of these focuses upon the analysis of sports skills to provide a better understanding of the execution of these movements so that teachers and coaches can work more effectively. In the past, progress was inhibited because of the lack of both qualified personnel and the necessary techniques to conduct the investigations. Fundamental research deals with the study of simple movements and factors which influence them. Strength, limb length, mass and inertial properties, angular and linear velocity and acceleration are examined. Results of this work form the basis for understanding the complexities of human motion and provide important insight into performance at the practical level. Theoretical research is relatively new to the field but will no doubt receive increasing emphasis in the future. It may involve the construction of a simplified mathematical model of a skill in a form which is suitable for computer analysis so that it can be simulated under several carefully controlled conditions. This leads to a thorough comprehension of the mechanics of the movement and has the potential for predicting more effective techniques. Research in the mechanical aspects of sport at the three levels described has increased sharply in complexity and sophistication. This has resulted from better instrumentation systems; utilization of knowledge from such fields as mechanics, engineering and computer science; and recognition of the contribution of biomechanics research to the improvement of performance in sport.

The authors have intended that this text bring together available information pertinent to the general field of biomechanics and be of special interest to persons in sport and physical education. Since the basis for the quantitative study of human movement is found in mechanics, it is essential that researchers be well prepared in statics and dynamics. These topics, along with the fundamentals of motion analysis and current information on body segment parameters, are reviewed in the earlier chapters. Instrumentation systems required for accurate measurement are described in the sections on photography and electronics. Computer techniques necessary for data processing, analysis and simulation are explained. The perplexing problem of locating relevant literature is also dealt with. Lastly, the fundamentals of biomechanics research are outlined and guidelines for the development of research and graduate programs are presented. The authors hope that this text, which represents their understanding of the current "state of the art," will be of value to those concerned with biomechanics in physical education and in related disciplines.

SELECTED REFERENCES

Alley, L. E.: Utilization of Mechanics in Physical Education and Athletics. JOHPER, *37,* 67–70, 1966 (March).

Barham, J. N.: Toward a Science and Discipline of Human Movement. JOHPER, *37,* 65–68, 1966. (October).

Braun, G. L.: Kinesiology: From Aristotle to the Twentieth Century. Res. Q. Amer. Assoc. Health Phys. Ed., *12,* 163–173, 1941.

Contini, R.: Preface. Hum. Factors, *5,* 423–425, 1963.

Contini, R., and Drillis, R.: Biomechanics. In *Applied Mechanics Surveys.* New York: Spartan Books, 1966a.

Contini, R., and Drillis, R.: Kinematic and Kinetic Techniques in Biomechanics. In F. Alt (Ed.), *Advances in Bioengineering and Instrumentation.* 1, New York: Plenum Press, 1966b.

Cooper, J. M. (Ed.): *Selected Topics on Biomechanics.* Chicago: Athletic Institute, 1971.

Evans, F. G.: Biomechanical Implications of Anatomy. In J. M. Cooper (Ed.), *Selected Topics on Biomechanics.* Chicago: Athletic Institute, 1971.

Fung, Y-C. B.: Biomechanics—Its Scope, History and Some Problems of Continuum Mechanics in Physiology. Appl. Mech. Rev., *21,* 1–20, 1968.

Groh, H.: Uber Entwicklungstendenzen der Biomechanik. Sportarzt und Sportmedizin, *8,* 351–354, 1968.

Hirt, S.: What is Kinesiology? A Historical Review. Phys. Therapy Rev., *35,* 419–426, 1955.

Kenedi, R. M.: Biomechanics in the Modern World. Biomed. Eng., *2,* 150–155, 164, 1967.

Lewillie, L., and Clarys, J. P. (Eds.): *Biomechanics in Swimming.* Brussels: Université Libre de Bruxelles, 1971.

Lissner, H. R.: Introduction to Biomechanics. Arch. Phys. Med. Rehab., *46,* 2–9, 1965.

Lissner, H. R.: Biomechanics—What Is It? Paper 62-WA-232 of the American Society of Mechanical Engineers, 1962.

Miller, D. I.: Computer Simulation of Human Motion. In H. T. A. Whiting (Ed.), *Techniques for the Analysis of Human Motion.* London: Henry Kimpton, In Press.

Nelson, R. C.: The New World of Biomechanics of Sport. Paper presented at the 75th Annual Convention of the National College Physical Education Association for Men, New Orleans, 1972.

Nelson, R. C.: Biomechanics of Sport: An Overview. In J. M. Cooper (Ed.), *Selected Topics on Biomechanics.* Chicago: Athletic Institute, 1971a.

Nelson, R. C.: Biomechanics of Sport: Emerging Discipline. Paper presented at the Third International Seminar on Biomechanics, Rome, 1971b.

Rasch, P. J.: Notes Toward a History of Kinesiology. Parts I, II and III, J. Amer. Osteopath. Ass., *58,* 572–574, 641–644, 713–714, 1958.

Steindler, A.: What has Biokinetics to Offer to the Physical Educator? Health Phys. Ed., *13,* 507–509, 555–556, 1942 (November).

Symposium Theorie der Sporttechnik. Warsaw: Akademie fur Korpererziehung, 1968.

Von Gierke, H. E.: Biodynamic Response of the Human Body. Appl. Mech. Rev., *17,* 951–958, 1964.

Vredenbregt, J., and Wartenweiler, J. (Eds.): *Biomechanics II.* Baltimore: University Park Press, 1971.

Wartenweiler, J., Jokl, E., and Hebbelinck, M. (Eds.): *Biomechanics.* Baltimore: University Park Press, 1968.

CHAPTER 2

Fundamentals of Analysis

JUST AS THE SUCCESSFUL PHYSICAL EDUCATOR must be proficient in the qualitative analysis of motion, the biomechanics researcher must be capable of making accurate quantitative evaluations of human performance. A systematic and theoretically sound approach similar to that used in classical mechanics can be adapted for the investigation of biomechanical problems. It should be recognized, however, that the human body is a far more complicated system than most encountered in the field of mechanics. Therefore, although the major emphasis is placed upon mechanical principles, appropriate modifications must be incorporated to account for the biological nature of man.

BIOMECHANICAL SYSTEMS

To view man purely as a machine is a gross oversimplification, but if the physical properties of the human body were represented exactly, mechanical analysis would be extremely difficult. The alternative is to construct a rather idealized model of the body in which such assumptions as the rigidity of individual segments are accepted. The purpose of developing these biomechanical systems is to achieve a practical combination of simplicity and accuracy which will facilitate quantitative analysis.

Within the context of the biomechanics of sport, the term "system" refers to one or more components which are capable of performing some common function related to human motion or sport. It may include the whole body, as in the case of a diver in free fall, or it may be limited to a few segments, depending upon the purpose of the analysis. In an investigation of kicking, the action of the segments of the leg contacting the ball might be of principal interest. In throwing, the system under consideration could be restricted to the trunk and throwing arm. The sports implement itself such as a golf club, hockey stick or discus might be the focus of the study. Whatever the problem, the boundaries of the system must be defined and the forces acting upon it identified.

For the purpose of mechanical analysis, the athlete may be treated as a particle, rigid body, quasi-rigid body or a linked system. When a system can be adequately represented by the motion of its mass center, the particle treatment may be used. Such is the case in predicting the trajectory of a jumper. Seldom is the athlete portrayed as a rigid body since this type of representation is better suited to the study of muscle and joint forces acting upon individual limb segments. Although the human body as a whole cannot be realistically considered as a single rigid unit, there are occasions when the segments remain in the same positions with respect to one another. This quasi-rigid configuration is seen in a diver rotating in a pike position or in a skater maintaining a fixed attitude in a spin, spiral or glide. A more accurate mechanical representation of the human body, however, is a series of interconnected rigid segments which demonstrate independent motion. The complexity of such a linked system is a function of the number of segments which it contains.

FORCES AND FREE BODY DIAGRAMS

Because any change in the state of rest or motion of a biomechanical system is governed by the action of external forces, it is important to identify these forces, their magnitudes, directions and points of application. To facilitate this process, a free body diagram should be constructed. All the external forces influencing the system are indicated on this free body diagram, which is a simplified drawing of the system isolated from its surroundings. To illustrate the principles of this approach to mechanical analysis, a brief discussion of the forces most commonly encountered in sport is presented along with selected examples and free body diagrams related to human motion.

Weight. The gravitational attraction which the earth exerts acts equally on all particles which compose a body. Weight, the resultant of this distributed force, is always located at the mass center of the body and is directed toward the center of the earth. Figure 2-1 shows the free body diagrams of a shot and an athlete in free fall in which air resistance is assumed to be negligible. In both instances, the arrow

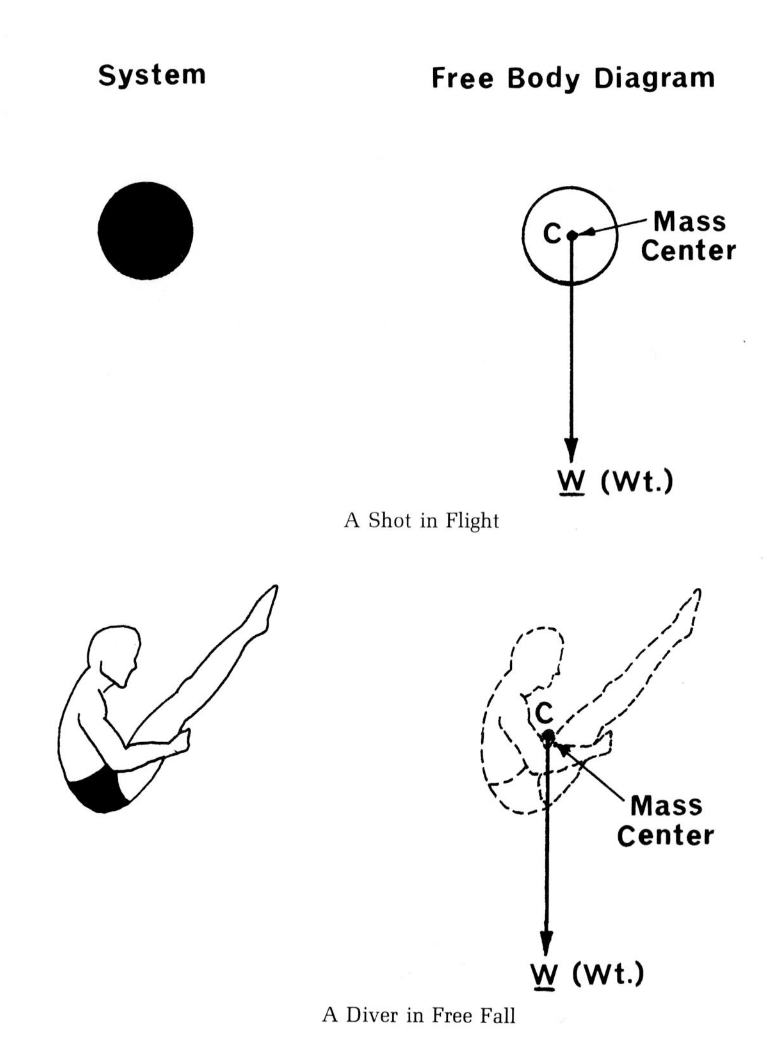

FIGURE 2-1. *Free Body Diagrams of Systems in Free Fall.*

indicates the direction and point of application of the weight. No other external force acts upon either system to influence its motion.

Reaction Forces. When a system is isolated in a free body diagram, forces of reaction replace all previous contacts of the system with its surroundings. A sprinter experiences such a reaction force from the ground when he thrusts his foot against it. Since force is a vector quantity having both magnitude and direction, it can be resolved into components whose vector sum is equal to the original force. It is convenient to represent a ground reaction force by two orthogonal components, one normal or at right angles and the other along or tangent to the surface. The tangential component, also known as the friction force, always opposes the direction of relative motion between the two surfaces. Since the foot tends to push downward and backward into the ground, the reaction of the ground is in the opposite direction (Figure 2-2) as a consequence of Newton's third law of motion (action and reaction). A similar situation exists with a shot putter just before he releases the shot. Since the system is defined as the athlete and the shot, the weights of both are shown acting vertically downward from their respective mass centers. The normal and tangential components of the ground reaction are indicated at the point of contact between the foot and the ground (Figure 2-2).

Letters depicting forces on free body diagrams are usually related to the force which they represent. Thus, **W** may stand for weight, **N** for the normal reaction and **F** or **Fr** for the friction force. Although such a convention is not essential, it does facilitate the rapid interpretation of the free body diagrams.†

Friction. Although several types of friction force exist, dry friction with its static and kinetic subdivisions is the most common in quantitative biomechanical analysis. Static friction implies a nonslip condition while kinetic friction is characterized by a sliding of the contact surfaces. As already stated, friction is the tangential component of the reaction force and always acts along the surface to oppose the relative motion or impending relative motion. The static (s) and kinetic (*k*) friction equations:‡

$$\mathbf{F}_{\mathbf{s}_{max}} = \mu_s * \mathbf{N} \qquad \mathbf{F}_{\mathbf{k}} = \mu_k * \mathbf{N}$$

indicate that the friction force **F** is a function of both the reaction normal to the surface, **N**, which represents the force pressing the

† Vectors are denoted by bold face type in the text and are underlined in the figures.

‡ To avoid confusion with the symbols for cross product (×) and dot product (·) used later, an asterisk (*) will be employed to indicate the operation of multiplication.

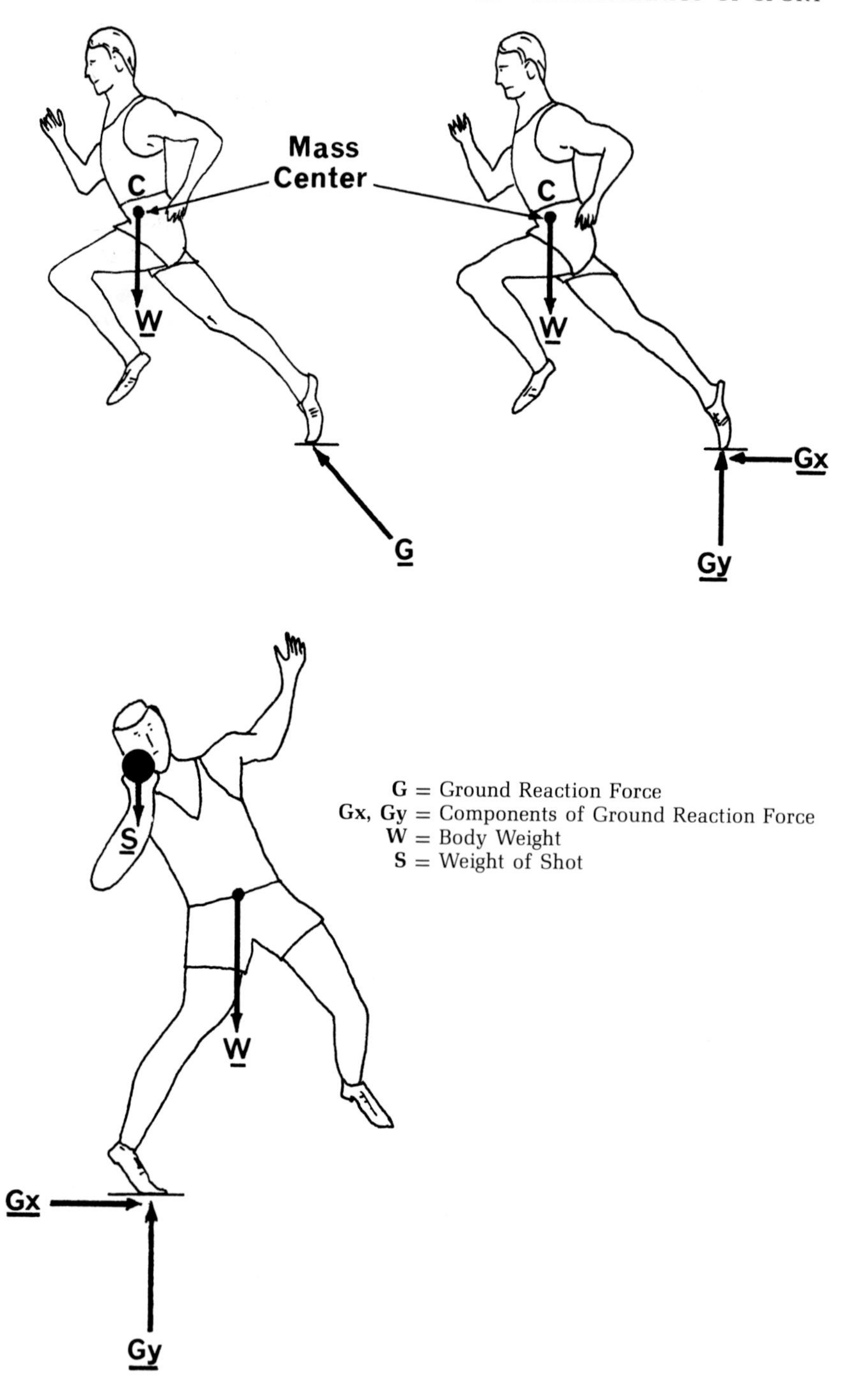

FIGURE 2-2. *Free Body Diagrams Indicating Reaction Forces.*

two surfaces together, and the coefficient of friction, μ, which depends upon the nature of the contacting surfaces. Dry, rough surfaces have higher coefficients of friction than those which are smooth and lubricated. Also, for a given pair of surfaces, the coefficient of static friction (μ_s) is slightly higher than the coefficient of kinetic friction (μ_k).

In the case of static friction, the equation specifies the maximum force which can be supported before slipping occurs. Special rubber-soled shoes and those with cleats or spikes worn by athletes increase the coefficient of friction and thus the maximum static friction force. As a result, the chance of slipping is less when running or changing direction. Grips on golf clubs and tennis racquets serve a similar purpose. It is important to realize that the static friction force will only reach its maximum value immediately before slipping occurs (Figure 2-3).

The kinetic friction force, commonly encountered in winter sports, is equal to the coefficient of kinetic friction multiplied by the normal force. In curling, friction between the rock and the ice is reduced by cleaning the sole of the rock prior to delivery and by sweeping to remove foreign material from the ice. Under fresh pebble conditions, sweeping also tends to polish the ice and provide a smoother path for the rock. Similarly, a minimum of resistance is desired in straight running in skiing (Outwater, 1970). Thus, the bases of skis are specially treated to reduce their coefficient of friction with the snow. For stopping, however, a large kinetic friction force must be built up to arrest the motion. Therefore, the sharp edges of the skis are pressed into the snow at an angle to the fall line.

In cross-country skiing, kinetic friction must be minimized when going downhill and static friction maximized when climbing (Outwater, 1970). A somewhat similar situation exists in skating, in which both kinetic and static friction forces are experienced. The sharpening of hockey, figure and speed skates provides well-defined edges to increase the kinetic friction employed in stopping and the static friction necessary for an effective thrust against the ice with the edge of the blade. Sharpening also removes the nicks and irregularities from the running surface so that a minimum of resistance is encountered in a direction parallel to the length of the blade. In skating, great pressure is applied to a small area of the ice, causing some of the energy to be dissipated as heat. This results in a slight melting beneath the blade and reduces the friction.

Muscle Force. Muscles, the internal motors responsible for generating purposeful human movements, exert a pull upon the

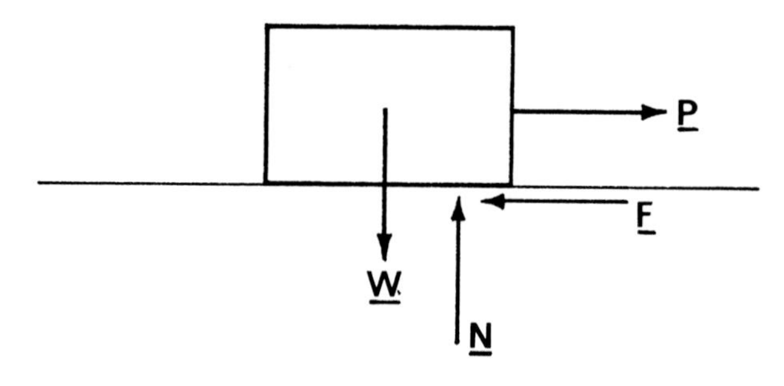

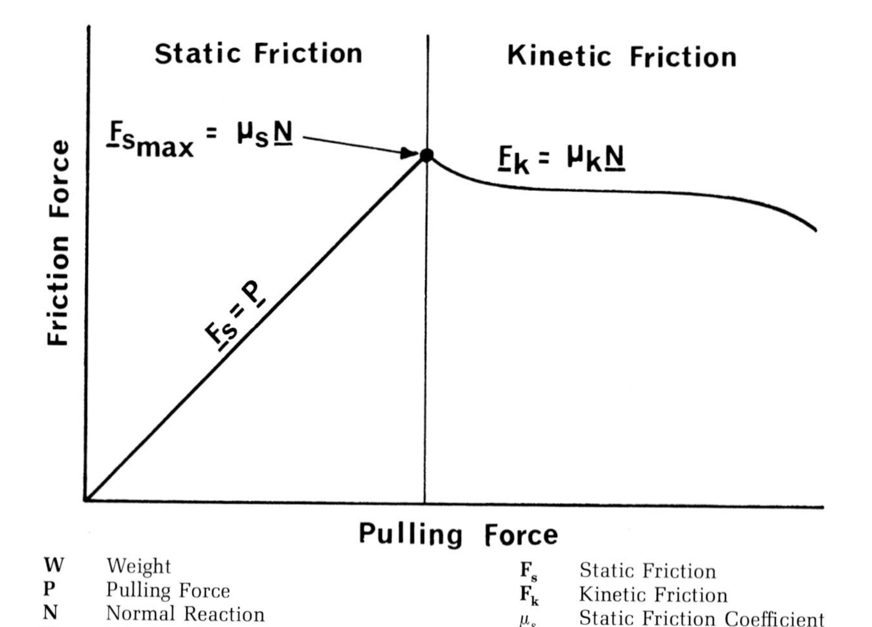

W Weight
P Pulling Force
N Normal Reaction
F Friction Force

F_s Static Friction
F_k Kinetic Friction
μ_s Static Friction Coefficient
μ_k Kinetic Friction Coefficient

FIGURE 2-3. *Friction Force Relationships* (*J. L. Meriam,* Statics. *New York, Wiley, 1966a*).

limbs to which they are attached. According to MacConaill and Basmajian (1969), the laws of approximation and detorsion dictate their potential actions. These laws state that when a muscle contracts, its bony attachments tend to be brought closer to one another and into the same plane. The actual forces and torques produced by individual muscles, however, cannot easily be predicted because of the indeterminate influence of a number of physiological and mechanical factors. These include length-tension and force-velocity relationships (Wilkie, 1968) as well as the location of the muscle attachments with respect to the joint. In addition, the axis of rotation of the joint may not remain fixed during the course of the movement. Therefore, it has become common practice to refer to the resultant torques or moments of force produced by all the muscles acting across a particular joint (Pearson *et al.*, 1963; Plagenhoef, 1968; Chaffin, 1969; and Dillman, 1971).

When a limb is considered separately from the rest of the body, the reaction force at the joint and the resultant muscle forces are classified as external and must be indicated on the free body diagram and included in the analysis. Figure 2-4 shows the upper arm and forearm-hand segments detached at the elbow and shoulder joints. The forces acting on the forearm at the elbow are equal in magnitude and opposite in direction to those acting upon the upper arm. If the elbow had remained intact, these would cancel one another and would not appear explicitly on the diagram. For the purpose of simplification, the resultant muscle force is shown as a single arrow

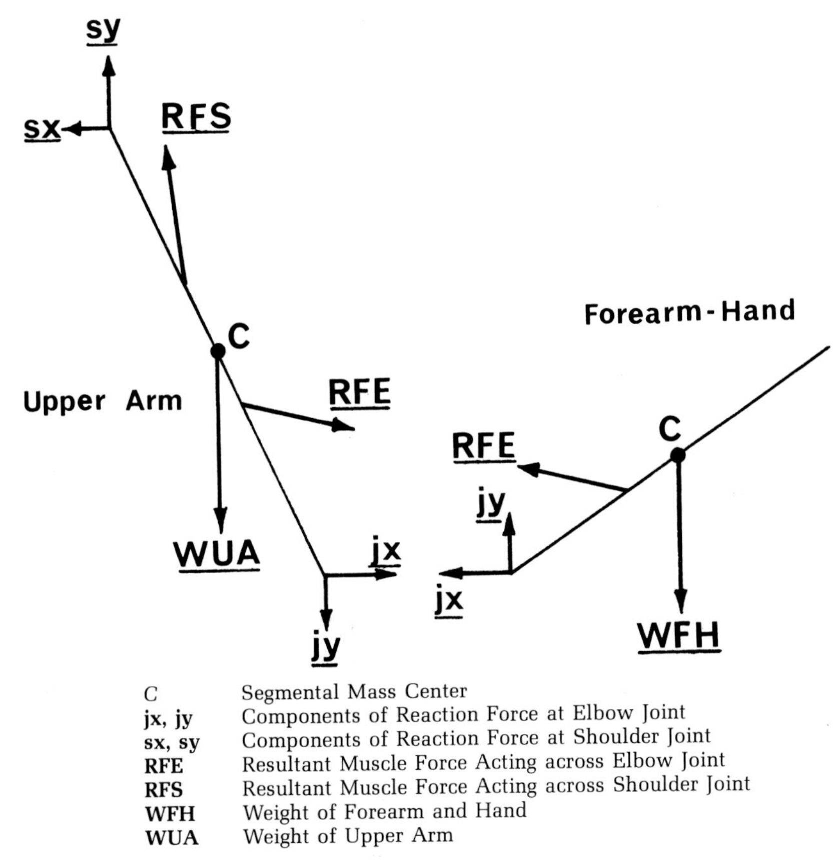

C	Segmental Mass Center
jx, jy	Components of Reaction Force at Elbow Joint
sx, sy	Components of Reaction Force at Shoulder Joint
RFE	Resultant Muscle Force Acting across Elbow Joint
RFS	Resultant Muscle Force Acting across Shoulder Joint
WFH	Weight of Forearm and Hand
WUA	Weight of Upper Arm

FIGURE 2-4. *Free Body Diagram of Isolated Body Segments.*

at an unknown distance from the joint. It could also be resolved into a stabilizing component tangent to the long axis and directed toward the joint, and a rotary component normal to the limb.

Impact. Impact forces result from the collision of two or more objects. In sport, they occur when striking a baseball with a bat or a golf ball with a club as well as in hitting the take-off board in a long jump or landing from a rebound. Athletic injuries caused by impact forces have recently become the focus of research (Hale, 1971). As a consequence, efforts are being made to develop adequate safety standards for protective sports equipment in the hope of reducing the number of fatalities and serious injuries suffered when a performer is struck by a pitched ball or hockey puck or when he runs into an immobile object such as a goal post. In the short time interval during which the collision takes place, extremely large impact forces are generated, so large in fact that all others are insignificant by comparison. On a free body diagram, an impact force is designated with an asterisk (*) beside the arrow indicating its line of action.

Buoyancy. Buoyancy is important in aquatic sports and particularly in survival skills. According to Archimedes' principle, the magnitude of the buoyant force is equal to the weight of the water which the body displaces. It is a distributed force represented in mechanical analysis by its resultant acting vertically upward through the center of buoyancy, which coincides with the center of volume of the body. A swimmer will experience a greater buoyant force as he submerges more and more of his body beneath the surface. Thus, the attempt of the panicky person to keep his arms, head and shoulders above the surface is not mechanically sound for in so doing he decreases the amount of water displaced and consequently reduces the magnitude of the buoyant force supporting him.

Fluid Forces. Competitive swimmers, skiers, speed skaters, cyclists and sky divers all experience significant fluid forces as do such projectiles as golf balls, baseballs, badminton birds and javelins. Because of the mathematical complexity of the field of fluid mechanics, however, exact solutions to this type of problem are extremely rare. For this reason, it is difficult to present a brief yet accurate account of fluid forces in sport. The reader desiring a complete discussion of this general area is, therefore, referred to current textbooks on fluid mechanics.

Although athletes and their projectiles generally move in still air (or water), for convenience of analysis and discussion, it is customary to consider them stationary within a fluid flow. The force exerted upon the body or object by the moving fluid can be resolved into two components which have physical significance to the analysis. The component parallel to the flow but in the opposite direction is referred to as drag. This includes both skin drag due to friction and form or pressure drag caused by stresses normal to the surface (Shames, 1962). The second component of force acts normal to the direction of the fluid flow and is termed lift. While the precise magnitudes of these components are difficult to determine, both are proportional to the product:

$$\rho A v^2$$

in which ρ is the fluid density, A represents the area of the body, and v is the velocity of the flow. Experimentally, investigations using wind tunnels have helped to provide objective evidence of the most efficient aerodynamic positions for skiers (Raine, 1970) and ski jumpers (Straumann, 1955; Tani and Iuchi, 1971). In addition, the effect of drag in various swimming positions has been estimated by measuring the force required to tow swimmers through the water (Kent and Atha, 1971).

SUMMARY

The forces most commonly encountered in the biomechanics of sport include those exerted by the weight, reaction of external objects, muscle tension, impact, buoyancy and fluids. Their effect upon a biomechanical system is best indicated on a free body diagram in which the system is considered isolated from its surroundings. Reaction forces are designated at points of contact with external surfaces and Newton's third law is applied to determine their directions. The weight of the system is always shown acting vertically downward from the mass center. Internal forces at intact joints do not appear on the free body diagram since they cancel one another. However, if the system is composed of one or more segments rather than the total body, then the reaction and resultant muscle forces at the exposed joints must be specified. The necessity to identify the external forces acting upon a biomechanical system and to represent them accurately on a free body diagram is stressed as this step lays the foundation for subsequent mechanical analysis involving the principles of statics and dynamics.

SELECTED REFERENCES

Alt, F. (Ed.): *Advances in Bioengineering and Instrumentation.* New York: Plenum Press, 1966.

Briggs, L. J.: Effect of Spin and Speed on the Lateral Deflection (Curve) of a Baseball; and the Magnus Effect for Smooth Spheres. Amer. J. Phys., *27,* 589–596, 1959.

Brown, R. M., and Counsilman, J. E.: The Role of Lift in Propelling the Swimmer. In J. M. Cooper (Ed.), *Selected Topics on Biomechanics.* Chicago: Athletic Institute, 1971.

Chaffin, D. B.: A Computerized Biomechanical Model—Development of and Use in Studying Gross Body Actions. J. Biomech., *2,* 429–441, 1969.

Cochran, A., and Stobbs, J.: *The Search for the Perfect Swing.* Philadelphia: Lippincott, 1968.

Counsilman, J. E.: *The Science of Swimming.* Englewood Cliffs, N.J.: Prentice-Hall, 1968.

Dempster, W. T.: Free-Body Diagrams as an Approach to the Mechanics of Human Posture and Motion. In F. G. Evans (Ed.), *Biomechanical Studies of the Musculo-Skeletal System.* Springfield, Ill.: C. C Thomas, 1961.

Dillman, C. J.: A Kinetic Analysis of the Recovery Leg During Sprint Running. In J. M. Cooper (Ed.), *Selected Topics on Biomechanics.* Chicago: Athletic Institute, 1971.

Dyson, G.: *The Mechanics of Athletics.* 5th Ed., London: University of London Press, 1970.

Farell, C.: Drag of Bodies Moving Through Fluids. In J. M. Cooper (Ed.), *Selected Topics on Biomechanics.* Chicago: Athletic Institute, 1971.

Francis, J. R. D.: *A Textbook of Fluid Mechanics.* 3rd Ed., London: Edward Arnold, 1969.

Hale, C. J.: Significant Trends and Complex Barriers in the Engineering of Protective Sports Equipment. Mater. Res. Stand., *11,* 8–12, 1971 (October).

Harrington, E. L.: An Experimental Study of the Motion of Curling Stones. Trans. Roy. Soc. Can., *18,* 247–259, 1924 (Third Series).

Kent, M. R., and Atha, J.: Selected Critical Transient Body Positions in Breast Stroke and Their Influence Upon Water Resistance. In L. Lewillie and J. P. Clarys (Eds.), *Biomechanics in Swimming.* Brussels: Université Libre de Bruxelles, 1971.

MacConaill, M. A., and Basmajian, J. V.: *Muscles and Movements—A Basis for Human Kinesiology.* Baltimore: Williams & Wilkins, 1969.

McMillan, R.: For Curlers Only. Roundel, *13,* 11–14, 1961 (January–February).

Meriam, J. L.: *Statics.* New York: Wiley, 1966a.

Meriam, J. L.: *Dynamics.* New York: Wiley, 1966b.

Outwater, J. O.: On the Friction of Skis. Med. Sci. Sports, *2,* 231–234, 1970.

Pearson, J. R., McGinley, D. R., and Butzel, L. M.: Dynamic Analysis of the Upper Extremity: Planar Motions. Hum. Factors, *5,* 59–70, 1963.

Plagenhoef, S.: Computer Programs for Obtaining Kinetic Data on Human Movement. J. Biomech., *1,* 221–234, 1968.

Raine, A. E.: Aerodynamics of Skiing. Sci. J., *6,* 26–30, 1970, (March).

Rogers, E. M.: *Physics for the Inquiring Mind.* Princeton: Princeton University Press, 1960.

Shames, I. H.: *Engineering Mechanics—Statics and Dynamics.* 2nd Ed., Englewood Cliffs, N.J.: Prentice-Hall, 1967.

Shames, I. H.: *Mechanics of Fluids.* New York: McGraw-Hill, 1962.

Straumann, R.: Vom Skisprung zum Skiflug. Sport (Zurich), *63,* 7–8, 1955 (May).

Tani, I., and Iuchi, M.: Flight-Mechanical Investigation of Ski Jumping. In The Society of Ski Science (Ed.), *Scientific Study of Skiing in Japan.* Tokyo: Hitachi, 1971.

Tricker, R. A. R., and Tricker, B. J. K.: *The Science of Movement.* New York: American Elsevier, 1967.

Wilkie, D. R.: *Muscle.* New York: St. Martin's Press, 1968.

Williams, M., and Lissner, H. R.: *Biomechanics of Human Motion.* Philadelphia: W. B. Saunders, 1962.

CHAPTER 3

Statics

STATICS is that branch of mechanics which is devoted to systems in equilibrium, that is, systems at rest or those undergoing non-accelerated motion. Under these circumstances, the sum of the external forces acting upon the system equals zero and there is an absence of rotation. A sound knowledge of the basic principles of statics provides a good foundation for the analysis of human motion and is necessary for the measurement of buoyancy, the determination of the location of the mass center of a body or system, and the design of research instrumentation to evaluate static strength. Forces influencing biomechanical systems are classified according to whether they act in a single plane (coplanar) or in several planes (non-coplanar). Within each of these two major classifications, they can be further categorized as parallel, concurrent (passing through a common point) or nonconcurrent nonparallel.

For coplanar, nonconcurrent, nonparallel force systems, there are three independent equilibrium equations, such as

$$\Sigma Fx = 0 \qquad \Sigma Fy = 0 \qquad \Sigma Mp = 0.$$

These relationships indicate that the sum of the forces acting upon the system in the x direction equals zero as does the sum of the forces in the y direction. In addition, the sum of the moments of force or torques about point p, which can lie anywhere in the xy plane, is also zero. A moment is a measure of the tendency of a force to produce rotation about a point or axis. It is equal to the product of the magnitude of the force and the perpendicular distance from the point or axis of rotation to the line of action of the force (moment arm). The direction of the moment is perpendicular to the plane containing the force and the moment arm. Its rotational sense is governed by the right hand rule (see Appendix B).

When dealing with noncoplanar forces, there are at most six independent equilibrium equations, and consequently, a maximum of six unknown forces or distances can be found. For example, the equations

$$\Sigma Fx = 0 \qquad \Sigma Fy = 0 \qquad \Sigma Fz = 0$$
$$\Sigma Mx = 0 \qquad \Sigma My = 0 \qquad \Sigma Mz = 0$$

show that the sum of the forces in three orthogonal directions and the sum of the moments about three mutually perpendicular axes are equal to zero. If the number of unknowns were to exceed the number of independent equations, the problem would be statically indeterminate and more information on the values of the forces and distances would be required before the equations could be completely solved.

MEASUREMENT OF BODY BUOYANCY

The determination of body buoyancy illustrates one case in which the principles of statics may be employed. If a swimmer is suspended by a spring scale (Figure 3-1), he is influenced by three external forces: **W**, the body weight acting vertically downward from the center of mass; **S**, the spring scale pulling upward; and **B**, the buoyant force equal to the weight of the water displaced by the swimmer and also directed upward. These forces are assumed to be parallel and coplanar. Since the system is in equilibrium, the sum of the forces equals zero.†

$$\Sigma Fy = 0$$
$$S + B - W = 0$$
$$B = W - S$$

†Throughout the book, the following mathematical conventions are observed: vectors directed upward or to the right are positive while those downward or to the left are negative. Counterclockwise rotations are considered positive while clockwise rotations are negative.

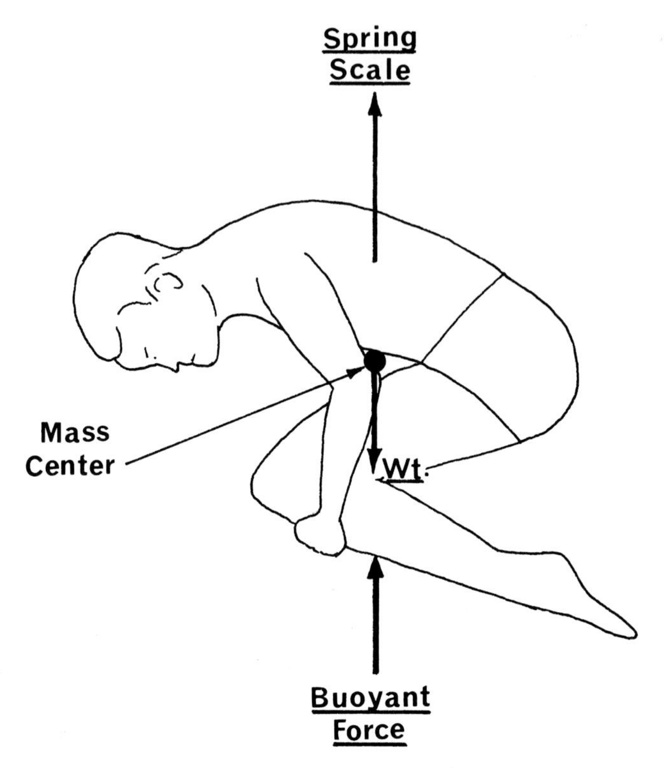

FIGURE 3-1. *Free Body Diagram Illustrating Buoyancy Measurement.*

If a 160-pound swimmer records a 3-pound force on the scale, the buoyant force which he experiences is equal to 157 pounds. Therefore, if the scale were removed, he would sink. By inhaling maximally, however, he might be able to increase his buoyant force to the point where he could float. Since inspiration expands the thorax, a greater volume of water is displaced without significantly increasing the body weight. Because the body settles in the water until it has displaced an amount of fluid equal to its weight, individuals float at different levels. The endomorph with his high ratio of body volume to body weight will float easily with his face, chest and toes above the surface. The mesomorph, however, is characterized by large proportions of bone and muscle whose density exceeds that of water. Therefore, he will have to be almost completely submerged before he is able to displace a volume of fluid whose weight equals that of his body. In addition, he will probably require conscious breath control to remain at the surface. The small minority of persons who, because of their body density, are unable to displace a sufficient amount of water even when completely immersed and maintaining maximum inspiration are termed "sinkers."

To complete the requirements for equilibrium, there cannot be any rotation of the body. Thus, the scale must be placed in such a way that the sum of the moments of the three forces, **S**, **B** and **W**, equals zero. If the scale is removed, this requirement can only be met when the lines of action of the buoyant force and the weight are collinear.

MASS CENTER LOCATION

The principles of statics may also be applied to locate the center of mass of the human body. The only apparatus required is a simple weight scale and a fairly sturdy board measuring approximately 2×7 feet with angle irons fastened to both ends. The board is positioned horizontally with one end resting upon the scale. A free body diagram of the board shown in Figure 3-2 indicates that the board weight **B** acts vertically downward from the center of mass which is located a distance *b* from the nonscale end. If the board were uniform and of homogeneous composition, its mass center and geometric center would be coincident. This assumption of uniformity, however, is not required to solve the problem at hand as will become evident in the calculations which follow. On the free body diagram, **Rx** and **Ry** are reaction force components exerted by the support upon the board. Similarly, **Sx** and **Sy** represent the horizontal and vertical reaction force components resulting from contact with the scale. The magnitude of **Sy** can be read directly from the scale.

Since the system is stationary, the equations of equilibrium apply. In this case, it is not necessary to determine all of the unknown forces and distances but only the moment or torque produced by the board. To obtain this information, moments are summed about the ZZ axis which coincides with the nonscale supporting edge. Since the lines of action of **Rx**, **Ry** and **Sx**, three of the reaction force components exerted upon the board by the supporting surfaces, pass through ZZ, each has a zero perpendicular distance to ZZ and, therefore, a zero moment with respect to it. This technique of eliminating unknown forces from the moment equation is a very useful one in mechanics. Because the moments can be summed about any axis, they are often taken about one which is intersected by the lines of action of forces or force components of unknown magnitude. A force whose line of action passes through or is parallel to an axis will not have any moment or turning effect about that axis.

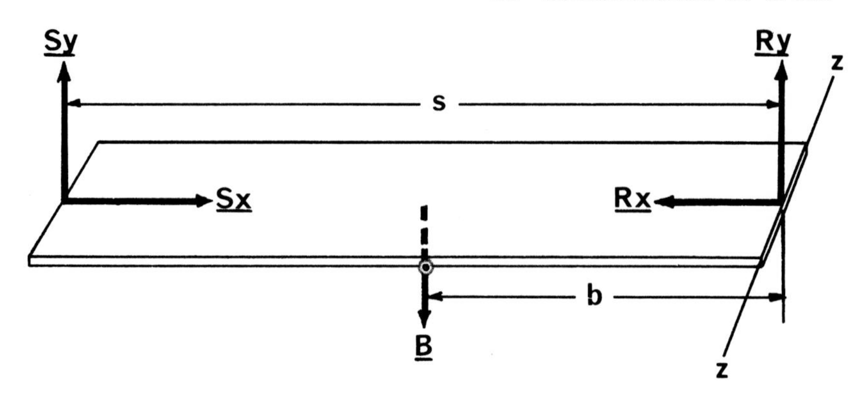

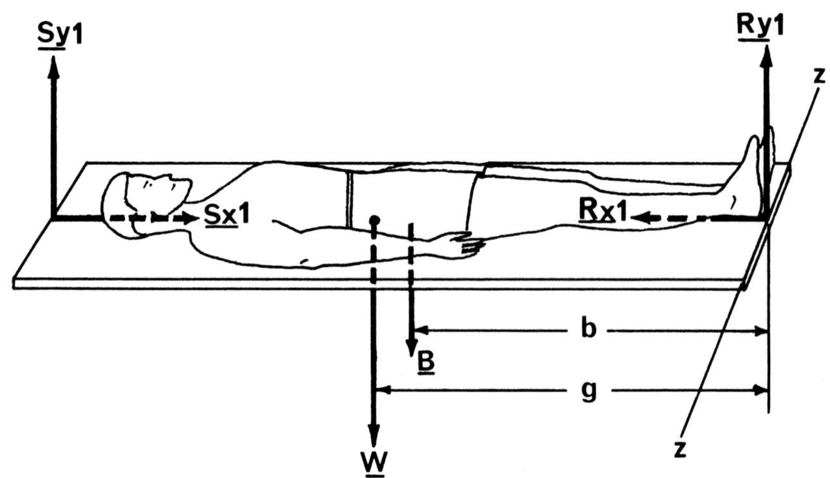

Sx, Sy	Reaction Force Components at the Scale End
Sx1, Sy1	
Rx, Ry	Reaction Force Components at the Nonscale End
Rx1, Ry1	
B	Weight of the Board
W	Weight of the Subject
s	Length of the Board between Supports
b	Distance from ZZ axis to the Line of Action of **B**
g	Distance from ZZ axis to the Line of Action of **W**

FIGURE 3-2. *Board and Scale Method for Determining Center of Mass.*

$$\Sigma Mzz = 0$$
$$-Sy * s + B * b = 0$$
$$B * b = Sy * s$$

If the scale reads 20 pounds and s, the distance between the two supporting edges of the board, is 80 inches, then the moment of the weight of the board ($B * b$) is 1600 pound-inches acting in a counterclockwise direction.

After this initial calculation, the subject lies supine on the board with his heels directly over the ZZ axis (Figure 3-2). The body weight **W** acts vertically downward from the mass center, which is located at an as yet undetermined distance g. Again moments are summed about the ZZ axis:

$$\Sigma Mzz = 0$$
$$-Sy1 * s + W * g + B * b = 0.$$

Assuming that the subject weighs 150 pounds and that the new scale reading $Sy1$ is 95 pounds:

$$-95 * 80 + 150 * g + 1600 = 0$$
$$150 * g = 7600 - 1600$$
$$g = 40 \text{ inches.}$$

This indicates that the mass center of the subject is located 40 inches from the ZZ axis. Since the subject's feet were placed directly over this axis, g also represents the location of the mass center with respect to the feet. This value can be expressed as a percentage of standing height. Thus, if the individual were 72 inches tall, his center of mass would be

$$\frac{40}{72} * 100 = 55.5\% \text{ of his standing height.}$$

While this application of statics provides an interesting laboratory exercise and is of academic interest, it is of limited value in the analysis of sports skills since the athlete seldom remains in one position for long.

MASS CENTER AS A FUNCTION OF BODY POSITION

A more utilitarian method of locating the mass center of the human body requires a knowledge of the masses and centers of mass of the segments.† This approach is based upon the concept of a resultant, namely, the simplest representation of a force system which can be made without altering the external effect of the system

†Refer to Chapter 5 for estimates of these segmental parameters.

upon a rigid body. The total body weight acting at the center of mass of the body is the resultant of the individual segment weights directed downward from their respective segmental mass centers. Thus, the body weight is equal to the sum of all the segment weights, and the rotation or moment which it can generate with respect to any given point is equivalent to the sum of the moments of the segmental weights about the same point.

To simplify the calculations initially, assume that a body of weight, **W**, is composed of only three segments: the head-trunk, which is 59.3% of the body weight; the legs, comprising 31.1% of the weight; and the arms, 9.6%. Figure 3-3 shows this idealized body in a rectangular coordinate system and indicates the x and y coordinates of its segmental mass centers. The determination of the x coordinate of the whole body involves calculating the moments of the segmental weights with respect to any point on the Y axis.

Moment of the Sum = Sum of the Moments

$$W * x = .593 * W * 10 + .311 * W * 15 + .096 * W * 13$$

Dividing both sides of the equation by W,

$$\begin{aligned} x &= .593 * 10 + .311 * 15 + .096 * 13 \\ &= 5.93 + 4.66 + 1.25 \\ &= 11.84. \end{aligned}$$

Thus, the x coordinate of the center of mass of the body is 11.84 in the coordinate system shown in Figure 3-3. A similar procedure is used to determine the y coordinate. In keeping with the concept of moments, the coordinate reference frame together with the diagrammatic representation of the body may be visualized as having been rotated 90 degrees so that the weight vectors are perpendicular to the Y axis. The y coordinate of each segmental mass center is then multiplied by the appropriate percentage of body weight and these values are summed.

$$\begin{aligned} y &= .593 * 20 + .311 * 5 + .096 * 25 \\ &= 11.86 + 1.56 + 2.4 \\ &= 15.82 \end{aligned}$$

The coordinates of the center of mass of the total body (11.84, 15.82) must subsequently be related to the positions of the body segments or to some external reference point of physical significance to the analysis. The numbers themselves have little meaning unless this is done.

This technique for determining the position of the mass center of the human body has many applications in biomechanical analysis. It can be used to locate the line of gravity in competitive swimming

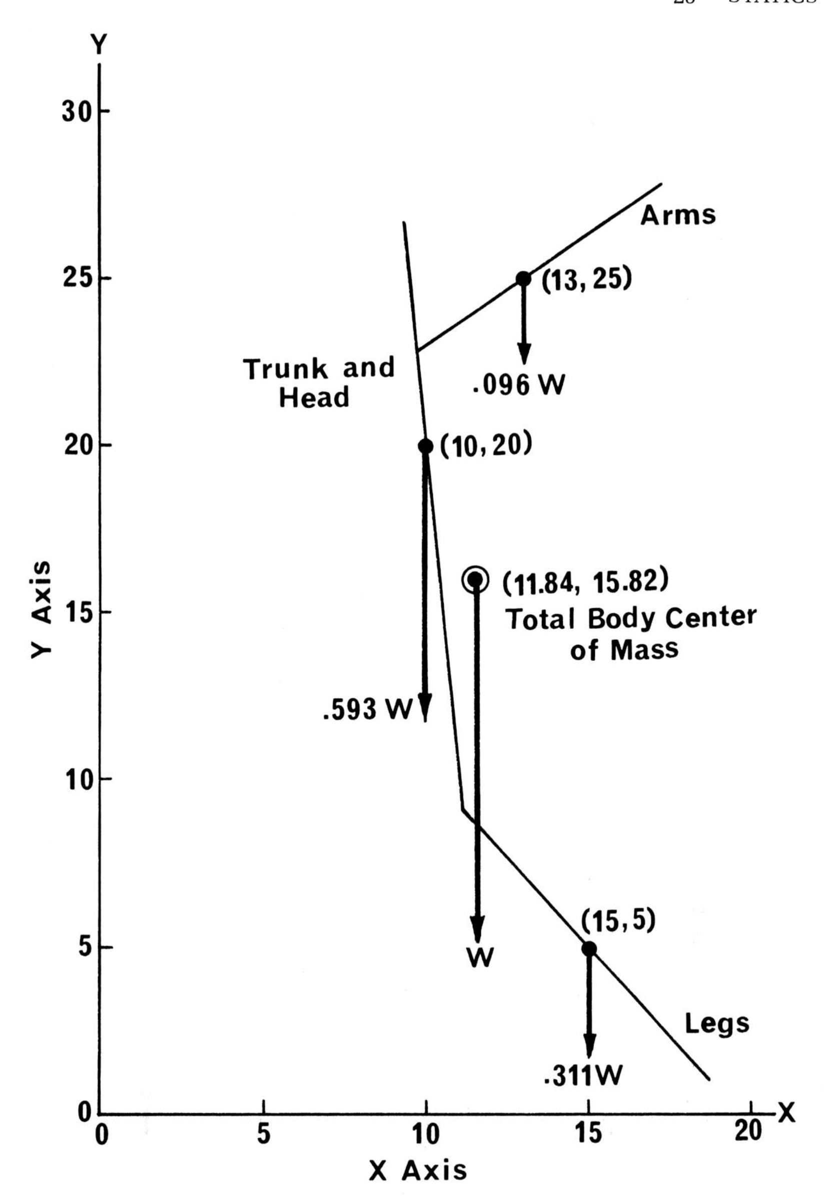

FIGURE 3-3. *Simplified Center of Mass Calculation.*

and track starts as well as in the various static poses in gymnastic free exercise and balance beam events. Such information is important in studying the relative stability of these positions. Since the mass center represents the body as a whole, its path during the course of a sports skill may be of particular interest. Does it pass under the bar in the high jump? How much is it displaced vertically in running? To what extent is it influenced by various changes in body position?

If the center of mass is to be located utilizing the method outlined, it is first necessary to have a picture of the athlete which has been or can be superimposed on a rectangular coordinate system. If segmental mass centers were not marked on the subject before the picture was taken, they will have to be calculated as functions of segment length. Suppose that the image of the forearm is two inches in length and its mass center is estimated to be 43% of the distance from the proximal end. Then, a ruler would be placed so that it was on a line connecting the joint centers of the wrist and elbow, and the mass center would be marked on the image of the limb a distance of .86 inch from the elbow. Similar procedures would be followed for the other segments.

While the mass centers can be located by hand as just described, it is faster and more efficient to use a template designed by Walton (1970) especially for this purpose (Figure 3-4). The appropriate portion of the template is simply laid over the representation of the

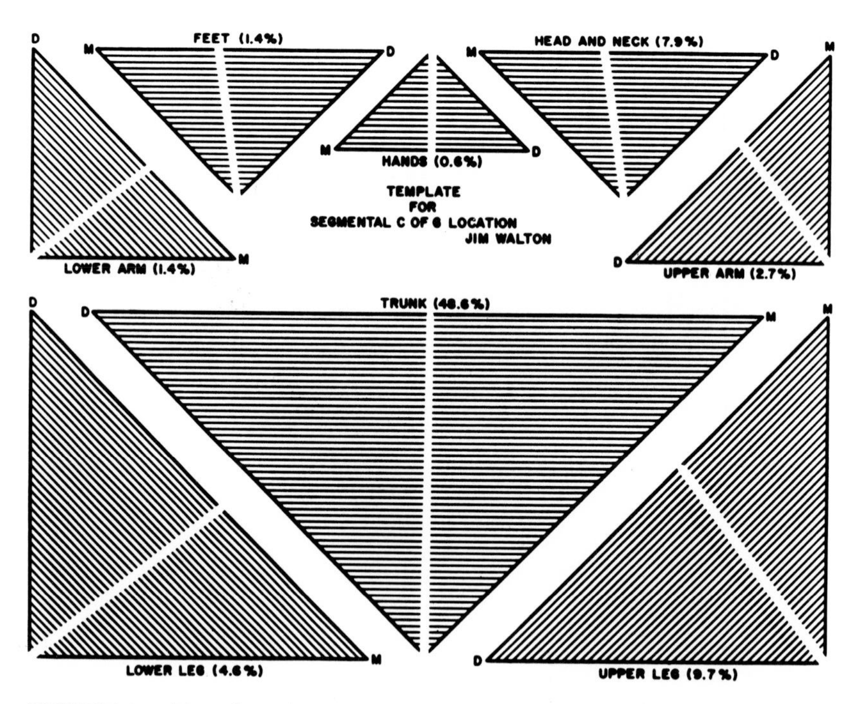

FIGURE 3-4. *Template for Determining Total Body Center of Mass* (J. S. Walton, *A Template for Locating Segmental Centers of Gravity. Research Quarterly of the American Association of Health, Physical Education, and Recreation 41:615–618, 1970*).

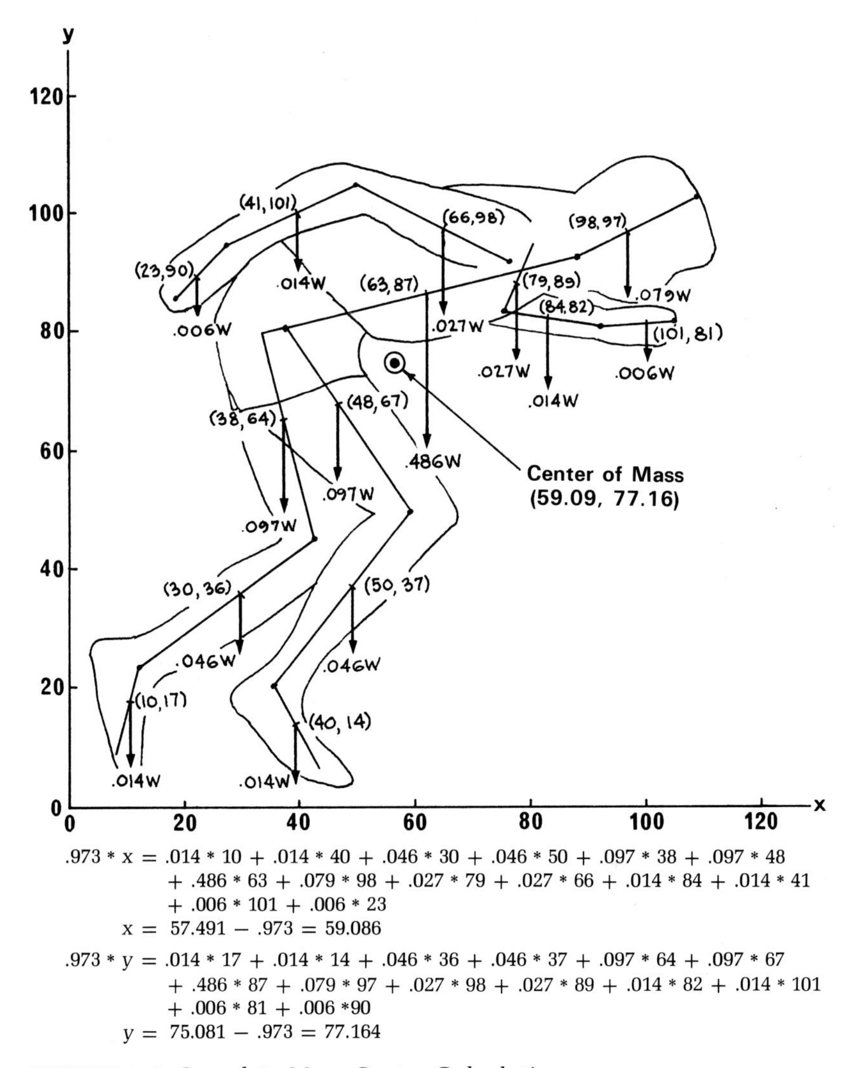

$$.973 * x = .014 * 10 + .014 * 40 + .046 * 30 + .046 * 50 + .097 * 38 + .097 * 48 + .486 * 63 + .079 * 98 + .027 * 79 + .027 * 66 + .014 * 84 + .014 * 41 + .006 * 101 + .006 * 23$$

$$x = 57.491 - .973 = 59.086$$

$$.973 * y = .014 * 17 + .014 * 14 + .046 * 36 + .046 * 37 + .097 * 64 + .097 * 67 + .486 * 87 + .079 * 97 + .027 * 98 + .027 * 89 + .014 * 82 + .014 * 101 + .006 * 81 + .006 * 90$$

$$y = 75.081 - .973 = 77.164$$

FIGURE 3-5. *Complete Mass Center Calculation.*

segment so that one of the lines coincides with the segment length. The center of mass is then approximated by the intersection of the segment line and the slot in the template.

Once the mass centers have been marked on the image, their coordinates can be determined with respect to the axes of the reference frame. The coordinates are then multiplied by their respective weight proportions and the products added (Figure 3-5). Division by .973 is necessary since the sum of the segmental weights on the template equals .973 of body weight. This discrepancy is due to blood and tissue losses in the original dissections.

The whole process of locating the mass center of the body from the position of the individual segments can be facilitated by employing a modern film analyzer with a built-in x-y coordinate system and by programming the necessary equations for solution on a digital computer.

MEASUREMENT OF STATIC STRENGTH

A knowledge of the principles of statics is required when evaluating the static strength of various muscle groups. Take, for example, the case in which a subject contracts his elbow flexors while his wrist is immobilized by a cuff. The free body diagram (Figure 3-6) shows the external forces acting upon the forearm-hand segment. The weight **W** of the segment is directed vertically downward from the mass center, which is located a distance s from the elbow joint. The magnitude of the force **T** exerted on the wrist by the cuff is

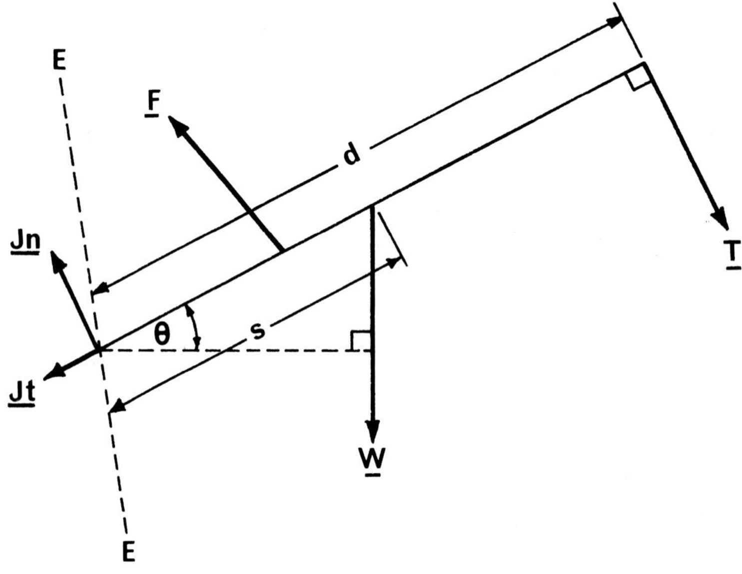

Jn, Jt	Normal and Tangential Components of Joint Reaction
EE	Axis of Rotation
W	Segmental Weight
T	Tension Exerted by Force Transducer
F	Resultant Muscle Force
θ	Angle of the Segment with the Horizontal
d	Distance from *EE* to the Point of Application of **T**
s	Perpendicular Distance from *EE* to the Line of Action of **W**

FIGURE 3-6. *Free Body Diagram of the Measurement of Elbow Flexion Static Strength.*

recorded by a force transducer. The transducer is positioned so that the line of action of the force is perpendicular to the long axis of the segment and its point of application is a distance *d* from the elbow joint. The resultant muscle force, **F**, which tends to produce elbow flexion, acts at an unknown distance from the joint. **Jn** and **Jt**, the normal and tangential components of the joint reaction force at the elbow, are present as a consequence of isolating the forearm-hand segment from the remainder of the body. From a knowledge of the values of the forces **T** and **W** and the distances s and *d*, the resultant muscle moment **RMM** can be computed. Since it is rarely possible to locate precisely the muscle insertions and angles of pull on a living subject, **RMM** represents the vector sum of all the muscle moments acting at the elbow joint.

Two slightly different methods can be used to calculate the magnitude of the resultant muscle moment. In the first, each force acting on the limb is resolved into two orthogonal components, one tangential or along the limb and the other at right angles or normal to the limb. According to the Principle of Moments, otherwise known as Varignon's theorem, the vector sum of the moments of the components of a force about a point is equal to the moment of the original force. Therefore, the moment of each component about the point of intersection of the horizontal axis *EE* through the elbow joint and the long axis of the limb can be calculated and the two moments summed. It will be observed (Figure 3-7), however, that a force component along the limb passes through the designated point to stabilize the joint. Therefore, it does not have any moment with respect to that point. Normal components acting at right angles to the segment account for the rotation of the forearm at the elbow.

$$\Sigma M_{Elbow\ joint} = 0$$
$$RMM - W * \cos\theta * s - T * d = 0$$
$$RMM = W * \cos\theta * s + T * d$$

The other technique requires that the perpendicular distance between the action line of the force and the axis be calculated. Since **T** is already acting at right angles to the limb, no further calculations are necessary. Distance *d* can be used as the moment arm of **T**. In the case of the weight **W**, the action line must be extended until a line can be drawn from the elbow axis to meet it at right angles. The moment arm can then be computed using the appropriate trigonometric function. In Figure 3-7, $\cos\theta = \frac{1}{s}$, therefore, the desired distance is equal to $s * \cos\theta$ and the magnitude of the moment of the weight is $W * s * \cos\theta$. Summing the moments

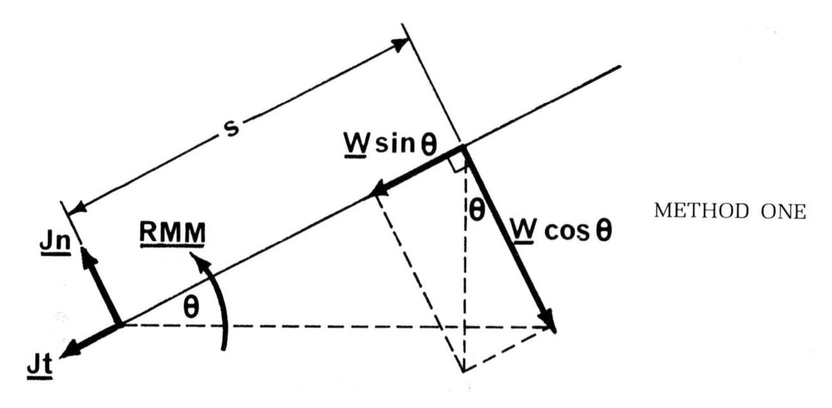

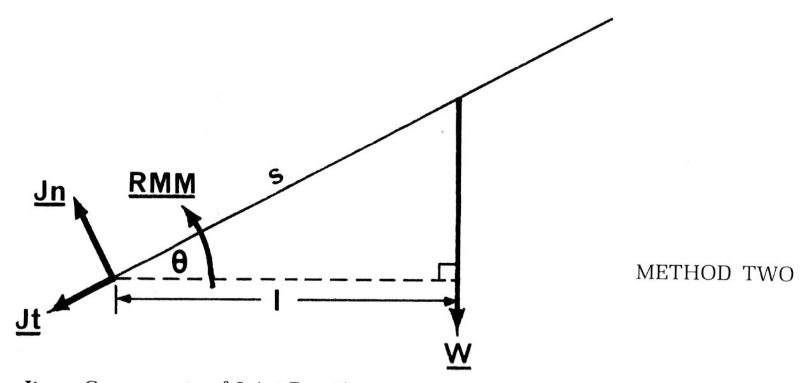

Jn, Jt	Components of Joint Reaction
θ	Angle of the Segment with the Horizontal
W	Segment Weight
RMM	Resultant Muscle Moment
s	Distance from Joint Center to Point of Application of **W**
l	Moment Arm of **W** with Respect to Joint Center

FIGURE 3-7. *Methods of Calculating the Moment of the Segmental Weight.*

of force with respect to the center of the elbow joint produces the magnitude of the resultant muscle moment:

$$\Sigma M_{Elbow\ joint} = 0$$

$$RMM - W * s * \cos\theta - T * d = 0$$

$$RMM = W * s * \cos\theta + T * d.$$

It is evident that both approaches produce the same result.

Once the magnitude of the resultant muscle moment has been determined for a specific position of the limb, the moment equation may be used to predict the amount of force T' which the forearm-hand segment can exert at any given distance d' from the elbow axis:

$$\Sigma M_{Elbow\ joint} = 0$$

$$RMM - W * s * \cos\theta - T' * d' = 0$$

$$T' = \frac{RMM - W * s * \cos\theta}{d'}$$

Since individuals vary in limb length, this calculation would be useful in comparing the flexion force which each could generate a common distance from the elbow.

It should also be borne in mind that the effect of segment weight is a function of the position of the forearm-hand with respect to the horizontal. As the arm approaches the horizontal and angle θ decreases, the force moment created by the weight approaches its maximum value. Assuming that the segment weighs 10 pounds and that the distance s to its mass center is 6 inches, the moment of the weight changes as follows for different angles:

Angle θ (Degrees)	Weight Moment ($s * W * \cos\theta$) (Pound-Inches)
0	$6 * 10 * \cos 0 = 6 * 10 * 1 = 60.00$
30	$6 * 10 * \cos 30 = 6 * 10 * .866 = 51.96$
45	$6 * 10 * \cos 45 = 6 * 10 * .707 = 42.42$
60	$6 * 10 * \cos 60 = 6 * 10 * .500 = 30.00$
90	$6 * 10 * \cos 90 = 6 * 10 * .000 = 0.00$

Therefore, the effective resistance caused by the weight is not constant if the arm is moved in a vertical plane. In some instances, this difficulty may be overcome by evaluating static strength in a horizontal plane.

HORIZONTAL ADDUCTIVE ARM STRENGTH

In the previous examples, the force systems considered have been predominantly coplanar. In biomechanics, however, there are many noncoplanar force systems which require three-dimensional solutions. The measurement of the horizontal adductive arm strength of the shoulder muscles falls into this category and provides a basic illustration of the vector analysis technique which may be applied in such circumstances.

In the free body diagram (Figure 3-8) of the right arm, forearm and hand in a rigid extended position, the force **T** applied to the arm by the force transducer is acting in the XY plane at right angles to the Y axis and a distance d from the shoulder joint. To simplify the analysis, the origin of the right-handed Cartesian system is arbitrarily placed at the shoulder joint with the Y axis coinciding

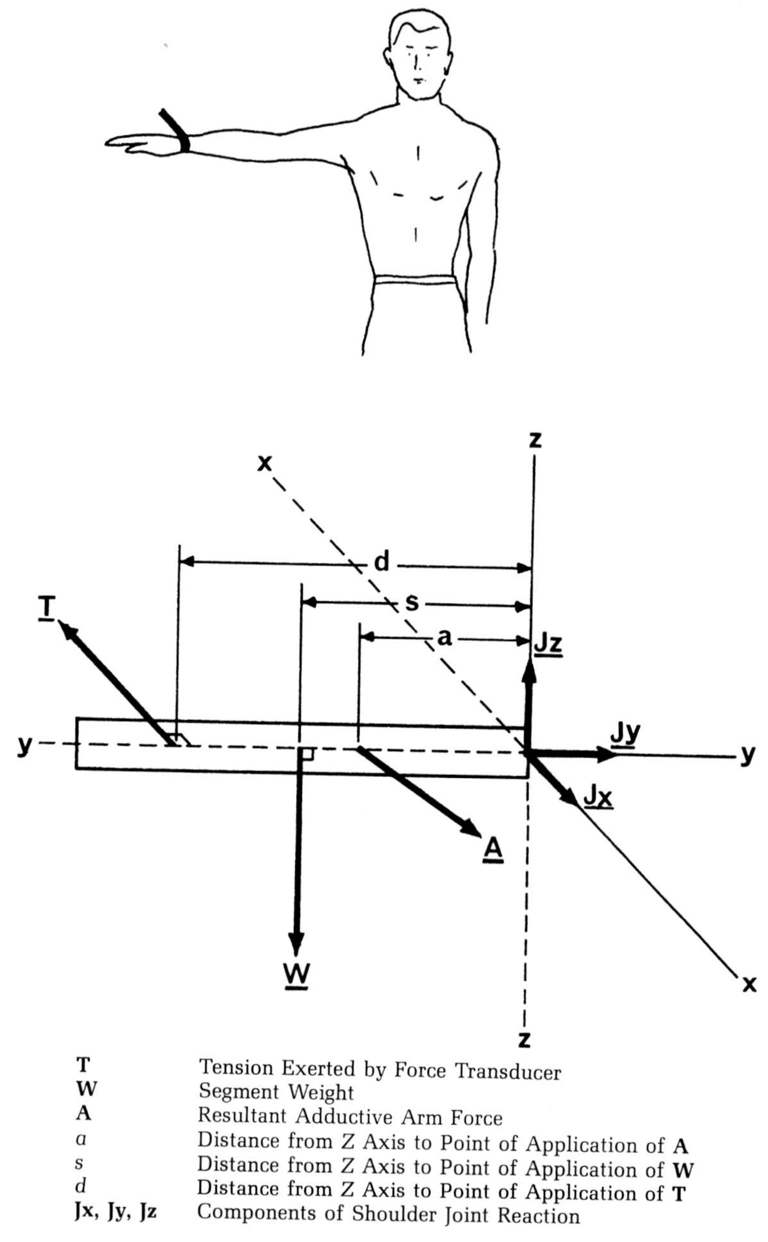

FIGURE 3-8. *Determination of Adductive Arm Strength.*

with the long axis of the arm. **W**, the weight of the limb segments, acts vertically downward in the *YZ* plane. The mass center of the system is a distance s from the shoulder joint. **Jx**, **Jy** and **Jz**, the components of the joint reaction along the X, Y and Z axes, are also indicated. **A**, the resultant adductive muscle force, is assumed to have its line of action in the XY plane.

Each of the forces is represented by the vector sum of its components in the three orthogonal directions. Thus,

$$\mathbf{J} = Jx\mathbf{i} + Jy\mathbf{j} + Jz\mathbf{k}$$
$$\mathbf{A} = Ax\mathbf{i} + Ay\mathbf{j}$$
$$\mathbf{W} = -W\mathbf{k}$$
$$\mathbf{T} = -T\mathbf{i}$$

in which **i**, **j** and **k** represent unit vectors directed along the X, Y and Z axes respectively.†

The triple scalar product may be utilized to determine the moments of these forces about the vertical Z axis through the shoulder joint. It is usually stated in general form as follows:

$$\mathbf{r} \times \mathbf{F} \cdot \mathbf{n}$$

in which **n** is the unit vector directed along the axis about which moments are to be taken. In this case, **n** is coincident with unit vector **k.** The designation **r** stands for a position vector joining any point on the axis represented by **n** with any point on the line of force vector **F**. Since both the magnitudes and directions of s and d are known, they serve as convenient position vectors for forces **W** and **T** respectively. Written in vector form, they are as follows:

$$\mathbf{s} = -s\mathbf{j}$$
$$\mathbf{d} = -d\mathbf{j}.$$

The composite adductive arm force **A** is assumed to produce a resultant muscle moment **RMM** about the Z axis.

Utilizing triple scalar products to sum the moments of force with respect to the Z axis:

$$\Sigma M_Z = 0$$
$$RMM + \mathbf{s} \times \mathbf{W} \cdot \mathbf{n} + \mathbf{d} \times \mathbf{T} \cdot \mathbf{n} = 0$$
$$RMM + \begin{vmatrix} 0 & -s & 0 \\ 0 & 0 & -W \\ 0 & 0 & 1 \end{vmatrix} + \begin{vmatrix} 0 & -d & 0 \\ -T & 0 & 0 \\ 0 & 0 & 1 \end{vmatrix} = 0$$
$$RMM + 0 - d * T = 0$$
$$RMM = d * T$$

The resultant muscle moment is equal in magnitude to the product of the force recorded by the transducer and the distance d. In this particular example, since forces with lines of action directed through

†A brief review of the fundamentals of vector algebra required for this section is presented in Appendix B. Shames (1967) is recommended for a comprehensive treatment of the topic.

or parallel to an axis have no turning effect or moment about that axis, the same result could have been obtained by summing the products of each of the forces and their respective perpendicular distances to the Z axis. Although the vector approach may not be necessary for all force problems encountered in biomechanics, it is particularly well suited for the analysis of complex noncoplanar systems.

In the past, the measurement of human strength has focused upon the maximum static or isometric force which an individual could exert utilizing various muscle groups. No limit was set upon the length of time required to achieve that maximum. In sport, however, the athlete seldom has unlimited time during which he can apply force. In the shot put, the velocity of the shot at release is the most important factor in determining the success of the put. Paradoxically, the greater the velocity, the less time the athlete is in contact with the shot. Likewise, as the runner increases his speed, he decreases the amount of time his foot is able to apply force against the ground. Therefore, from a practical standpoint, it would appear more important for a performer to be able to build up force quickly rather than to achieve a higher absolute force in the absence of any time restrictions. In response to this logic, several researchers including Clarke (1964), Royce (1962) and Sukop and Reisenauer (1968) have examined the nature of the isometric force-time relationship (Figure

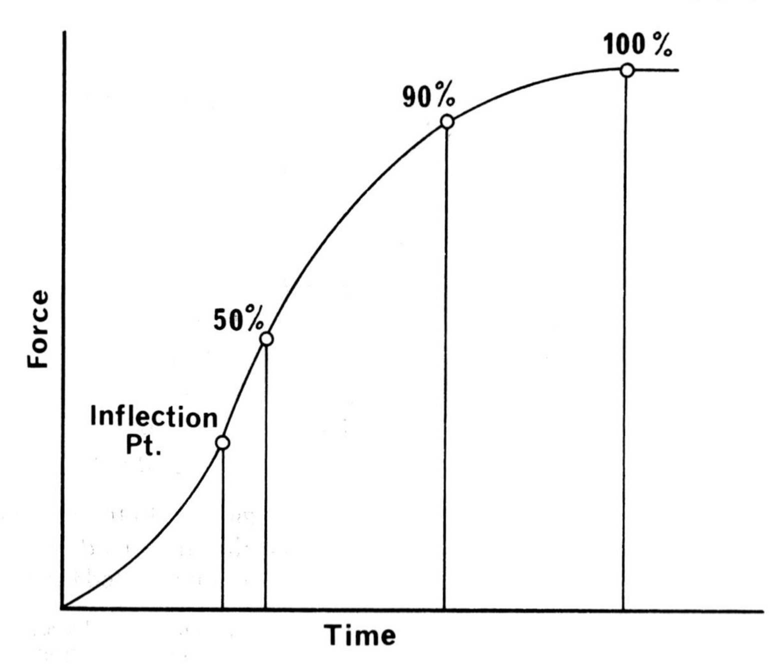

FIGURE 3-9. *Static Force-Time Relationship.*

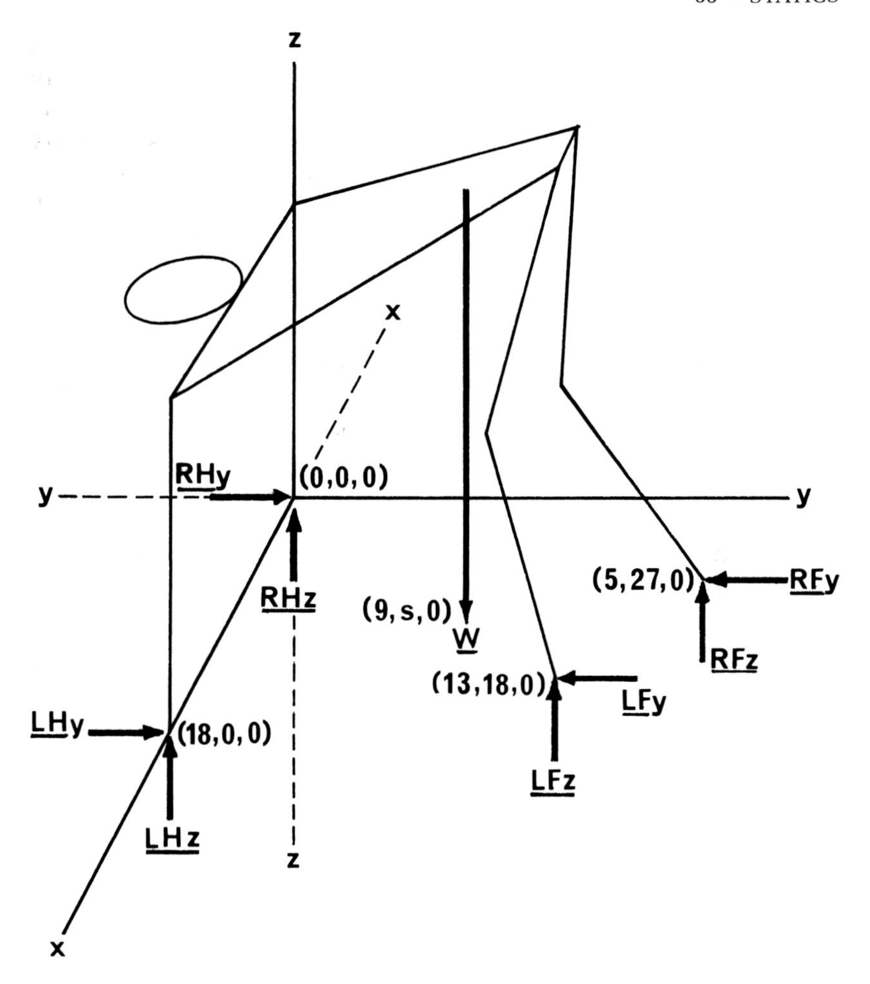

FIGURE 3-10. *Free Body Diagram of a Sprint Start Set Position.*

3-9). Utilizing procedures similar to those described for the measurement of static strength, these investigators have made continuous recordings of the force development of various muscle groups to study the influence of such factors as age, sex, fatigue, training and diurnal changes upon the curves.

SPRINT START

A runner in the "set" position of a sprint start provides an example of a biomechanical system in which the external forces are nonconcurrent and noncoplanar. Since the athlete is in equilibrium, the principles of statics may be employed in the analysis. In the free body diagram of the runner (Figure 3-10), reaction forces replace ground contacts with the hands and feet. Assuming that the start is performed correctly, the components of the ground reactions are limited to the Y and Z directions, with the X components being considered zero or negligible. The body weight acts vertically downward from the center of mass. Figure 3-10 indicates the positions of the hands and feet in a right-handed Cartesian coordinate system which has its origin coincident with the right hand and its X axis directed along the line joining the hands. The coordinates of the right hand (0, 0, 0), left hand (18, 0, 0), right foot (5, 27, 0) and left foot (13, 18, 0) are given in inches. These coordinates represent the points of application of the resultant of each ground reaction force acting upon its respective extremity.

Since the forces act in parallel planes, there are only five independent equations of equilibrium:

$$\Sigma Fy = 0$$
$$LHy + RHy - RFy - LFy = 0 \qquad (1)$$

$$\Sigma Fz = 0$$
$$RHz + LHz + RFz + LFz - W = 0 \qquad (2)$$

$$\Sigma Mxx = 0$$
$$\begin{vmatrix} 0 & 18 & 0 \\ 0 & -LFy & LFz \\ 1 & 0 & 0 \end{vmatrix} + \begin{vmatrix} 0 & 27 & 0 \\ 0 & -RFy & RFz \\ 1 & 0 & 0 \end{vmatrix} + \begin{vmatrix} 0 & s & 0 \\ 0 & 0 & -W \\ 1 & 0 & 0 \end{vmatrix} = 0$$
$$18\,LFz + 27\,RFz - s * W = 0 \qquad (3)$$

$$\Sigma Myy = 0$$
$$\begin{vmatrix} 18 & 0 & 0 \\ 0 & LHy & LHz \\ 0 & 1 & 0 \end{vmatrix} + \begin{vmatrix} 13 & 0 & 0 \\ 0 & -LFy & LFz \\ 0 & 1 & 0 \end{vmatrix} + \begin{vmatrix} 5 & 0 & 0 \\ 0 & -RFy & RFz \\ 0 & 1 & 0 \end{vmatrix} + \begin{vmatrix} 9 & 0 & 0 \\ 0 & 0 & -W \\ 0 & 1 & 0 \end{vmatrix} = 0$$
$$-18\,LHz - 13\,LFz - 5\,RFz + 9\,W = 0 \qquad (4)$$

$$\Sigma Mzz = 0$$
$$\begin{vmatrix} 18 & 0 & 0 \\ 0 & LHy & LHz \\ 0 & 0 & 1 \end{vmatrix} + \begin{vmatrix} 13 & 0 & 0 \\ 0 & -LFy & LFz \\ 0 & 0 & 1 \end{vmatrix} + \begin{vmatrix} 5 & 0 & 0 \\ 0 & -RFy & RFz \\ 0 & 0 & 1 \end{vmatrix} = 0$$
$$18\,LHy - 13\,LFy - 5\,RFy = 0 \qquad (5)$$

Since this particular case is quite straightforward, the moment equations could have been determined by summing the products of the forces and their respective perpendicular distances to the Z axis.

An examination of the equilibrium equations reveals that the system is statically indeterminate. Even if the value of the weight is given, nine unknowns (*LHy*, *LHz*, *RHy*, *RHz*, *LFy*, *LFz*, *RFy*, *RFz* and s) remain. Since a maximum of five unknowns can be determined from the five independent equations, it is evident that additional information is required before the equations can be completely solved. If triaxial force plates were attached to the starting blocks, *RFy*, *RFz*, *LFy* and *LFz* could be obtained experimentally. Substitution of these values into the equations would permit the calculation of the remaining unknowns. For laboratory purposes, the problem could be reduced to one of estimating the location of the line of gravity with respect to the hands in various modifications of the "set" position. A weigh scale placed under each foot would indicate *RFz* and *LFz*. After making appropriate measurements, moments of force could be summed about the X axis to calculate s, the distance to the line of gravity.

Static indeterminacy is encountered in many biomechanical problems. Because reaction force components are difficult to determine, the number of unknowns often exceeds the number of equations. Even though it may not be possible to solve for all the unknowns, some of the values may be determined by carefully selecting the axes about which to sum the moments of force.

SUMMARY

The importance of constructing an accurate free body diagram before attempting to solve any problem in statics cannot be overemphasized. All external forces acting upon the system must be included. These are often expressed in terms of their components acting along mutually perpendicular axes. If the system is coplanar, there are, at most, three independent equations for determining unknown forces and distances. In noncoplanar, nonparallel systems, three force and three moment equations make it possible to solve for a maximum of six unknowns. The careful selection of appropriate axes about which to calculate moments of force is essential and may spell the difference between success and failure in solving equilibrium equations. An axis through which several unknown forces act will considerably reduce and simplify the moment relationships. The method of multiplying the force by the perpendicular distance to calculate moments employed in coplanar analyses may also be applied to the simpler noncoplanar force systems. More complex noncoplanar systems, however, will be more efficiently dealt with by using the vector approach.

SELECTED REFERENCES

Butler, A. J.: Racial Comparison of Isometric Force-Time Characteristics of College Women. Unpublished Master's Thesis, Pennsylvania State University, 1970.

Chaffin, D. B.: A Computerized Biomechanical Model—Development of and Use in Studying Gross Body Actions. J. Biomech., *2*, 429–441, 1969.

Clarke, D. H.: The Correlation between Strength and the Rate of Tension Development of a Static Muscular Contraction. Int. Z. Angew. Physiol., *20*, 202–206, 1964.

Meriam, J. L.: *Statics*. New York: Wiley, 1966.

Royce, J.: Force-Time Characteristics of the Exertion and Release of Hand Grip Strength under Normal and Fatigued Conditions. Res. Q. Amer. Assoc. Health Phys. Ed., *33*, 444–450, 1962.

Shames, I. H.: *Engineering Mechanics—Statics and Dynamics*. 2nd Ed., Englewood Cliffs, N.J.: Prentice-Hall, 1967.

Stothart, J. P.: Relationships between Selected Biomechanical Parameters of Static and Dynamic Muscle Performance. Paper presented at the Third International Seminar on Biomechanics, Rome, 1971.

Sukop, J., and Reisenauer, R.: The Changes in Muscular Contraction and Relaxation after the Static Load in 16-Year Boys and Girls. Cesk. Hyg., *13*, 458–466, 1968.

Walton, J. S.: A Template for Locating Segmental Centers of Gravity. Res. Q. Amer. Assoc. Health Phys. Ed., *41*, 615–618, 1970.

Willems, E. J.: Het Verband Tussen de Kracht en de Snelheid bij een Willekeurige Maximale Isometrische Spieramentrekking. Unpublished Doctoral Dissertation, University of Leuven, Belgium, 1970.

Williams, M., and Lissner, H. R.: *Biomechanics of Human Motion*. Philadelphia: W. B. Saunders, 1962.

CHAPTER 4

Dynamics

IN THE BIOMECHANICAL FRAME OF REFERENCE, dynamics is concerned with man in motion and may also be extended to include implements which he manipulates or projects. From the standpoint of dynamics, sports skills may be examined on three levels. The first is a temporal analysis which deals with the timing or rhythm of various aspects of performance. The second, a kinematic investigation, concentrates upon the geometry of the motion without regard to the forces producing it. Displacement, velocity and acceleration are included at this level. Finally, a kinetic analysis is devoted to a study of the forces initiating, altering and stopping the motion. Kinetics is the most detailed of the three levels and requires the greatest understanding of fundamental mechanics.

TEMPORAL ANALYSIS

One of the first steps in investigating the mechanical basis of a motor skill is to gain an appreciation of the timing sequence of the movement components. In running, for example, a temporal analysis includes the determination of the time of support and the time of flight; in a gymnastics vault, the duration of the pre-flight, contact with the horse and after-flight; and, in swimming, the time required for the recovery, catch and propulsive phases of the stroke. Almost any sports skill can be subdivided into similar components which have practical significance when they are related to the performance. This type of analysis can be readily used in conjunction with cinematography since the number of frames of film required to complete a given phase of the action can be converted to elapsed time. Each portion of the skill must be carefully identified to ensure accuracy and reliability of the data. Precise operational definitions of the initiation and completion of the particular movement components under consideration are also required.

Temporal analysis may be utilized to investigate several questions of concern to both the practitioner of the sport and the researcher. How is the timing of the skill influenced by the speed of the performance? By fatigue? Body size? How great are the intra-individual differences in highly skilled performers? In novices? How do they change with learning and practice? Do changes in athletic equipment or sports implements affect the overall rhythm of the skill? Consideration of these and related problems make it evident that this level of investigation is of value to the teacher and coach. In addition, it provides a sound basis for subsequent kinematic and kinetic analysis of motion.

KINEMATICS

Kinematics, the geometry of motion, is that branch of dynamics which deals with displacement (change in position), velocity and acceleration. No reference, however, is made to the forces responsible for the motion. A summary of the symbols and basic kinematic relationships is presented in Table 4-1. Information pertaining to a particular sports skill or human movement must be translated into such terms if the methods of mechanics are to be applied to the analysis.

While similar principles are employed in calculating both the linear and angular parameters, the researcher should realize that linear displacement, velocity and acceleration refer to the motion of a particle or a specific point on a rigid body†. In contrast, angular displacement, velocity and acceleration are the same for all lines on a rigid body. Thus, they apply to the rigid body as a complete unit. It should also be noted that a particle cannot experience angular motion because of its negligible mass.

When velocity or acceleration is determined over a time interval,

†An exception to this general principle will be noted in a subsequent discussion of translation.

Table 4-1. Kinematics Summary

Parameter	*Linear*	*Angular*
Position—location with respect to a particular reference frame	**r** (x, y, z) **r** is a position vector between the origin of the coordinate system and the particular point. It is usually expressed as: $\mathbf{r} = x\mathbf{i} + y\mathbf{j} + z\mathbf{k}$ (inches, feet, meters, . . .)	θ (degrees, radians, revolutions, . . .)
Displacement—change in position	$\mathbf{s} = \Delta\mathbf{r}$ (inches, feet, meters, . . .)	$\Delta\boldsymbol{\theta}$ (degrees, radians, revolutions, . . .)
Velocity—change in position with respect to time	$\mathbf{v}_{ave} = \dfrac{\Delta\mathbf{r}}{\Delta t}$ $\quad \mathbf{v} = \dfrac{d\mathbf{r}}{dt}$ (ft/sec, mph, m/sec, . . .)	$\omega_{ave} = \dfrac{\Delta\boldsymbol{\theta}}{\Delta t}$ $\quad \omega = \dfrac{d\boldsymbol{\theta}}{dt}$ (deg/sec, rad/sec, rpm, . . .)
Acceleration—change in velocity with respect to time	$\mathbf{a}_{ave} = \dfrac{\Delta\mathbf{v}}{\Delta t}$ $\quad \mathbf{a} = \dfrac{d\mathbf{v}}{dt}$ (ft/sec/sec, m/sec/sec, . . .)	$\boldsymbol{\alpha}_{ave} = \dfrac{\Delta\omega}{\Delta t}$ $\quad \boldsymbol{\alpha} = \dfrac{d\omega}{dt}$ (deg/sec/sec, rad/sec/sec, . . .)

the result obtained is an "average" which is assumed to occur at the midpoint of the time interval. Thus,

$$\mathbf{v} = \frac{\Delta\mathbf{r}}{\Delta t} \quad \text{and} \quad \omega = \frac{\Delta\boldsymbol{\theta}}{\Delta t}$$

represent "average" linear and angular velocities respectively while

$$\mathbf{a} = \frac{\Delta\mathbf{v}}{\Delta t} \quad \text{and} \quad \boldsymbol{\alpha} = \frac{\Delta\omega}{\Delta t}$$

indicate the corresponding accelerations. As an example, consider the final stages of a hypothetical sprint in which the horizontal motion of the athlete's mass center is examined (Table 4-2). The Δt over which the successive horizontal velocities and accelerations are calculated is one second up to $t = 10$. Thus, the "average" velocity during the period between 9 and 10 seconds is 35 feet per second effective at $t = 9.5$. At the completion of the race, however, the final velocity (33 ft/sec) is calculated over a .6 second time interval, the

Table 4-2. Horizontal Motion of the Mass Center During a Sprint

Time (sec)	*Position* x† (ft)	Δx (ft)	*Velocity* v_x (ft/sec)	Δv_x (ft/sec)	*Acceleration* a_x (ft/sec/sec)
5	109				
		32	32		
6	141			2	2
		34	34		
7	175			1	1
		35	35		
8	210			0	0
		35	35		
9	245			0	0
		35	35		
10	280			−2	−2.5
		20	33		
10.6	300				

†Horizontal position of the mass center with respect to the starting line.

midpoint of which occurs at $t = 10.3$ seconds. Therefore, the final acceleration is

$$a_x = \frac{\Delta v_x}{\Delta t} = \frac{33 - 35}{10.3 - 9.5} = -2.5 \text{ ft/sec/sec.}$$

Velocities and accelerations can also be "instantaneous." This condition is met when the time interval over which they are calculated is extremely small and, by strict definition, approaches zero. Therefore,

$$\mathbf{v} = \lim_{\Delta t \to 0} \frac{\Delta \mathbf{r}}{\Delta t} = \frac{d\mathbf{r}}{dt} \qquad \boldsymbol{\omega} = \lim_{\Delta t \to 0} \frac{\Delta \boldsymbol{\theta}}{\Delta t} = \frac{d\boldsymbol{\theta}}{dt}$$

and

$$\mathbf{a} = \lim_{\Delta t \to 0} \frac{\Delta \mathbf{v}}{\Delta t} = \frac{d\mathbf{v}}{dt} \qquad \boldsymbol{\alpha} = \lim_{\Delta t \to 0} \frac{\Delta \boldsymbol{\omega}}{\Delta t} = \frac{d\boldsymbol{\omega}}{dt}$$

represent instantaneous values which occur at the instant they are calculated.

Classification of Motion. Human movement may be classified as translation, rotation, general plane motion or general space motion. The latter represents the general case and the first three categories are simply special cases of general space motion. Translation refers to movement in which there is no angular displacement of the body during any time interval. Thus, all parts of the body have the same linear acceleration. Rotation implies that particles of a rigid body follow circular paths about a fixed point. A combination of translation and rotation in a single plane is termed general plane motion. It is rotation about a point which itself is moving. When the latter occurs in more than one plane, it is known as general space motion. Most human movement incorporates both translation and rotation. Sports skills such as the golf swing, the tumble turn in swimming, the discus throw and the twisting somersault are examples of general space motion. Their analysis is a complex task. With other skills including running, long jumping, kicking and gymnastic vaulting, movement in the third dimension is often assumed negligible and they are treated as cases of general plane motion. This considerably simplifies the mechanical analysis.

In biomechanics, reference to the translation of the human body usually indicates movement of the mass center. Thus, the take-off of a jumper or the horizontal acceleration of a sprinter is calculated from the changing positions of his center of mass over specified time periods. In these instances, the athlete is treated mathematically as a particle and it is assumed that his motion can be described by studying the kinematics of his mass center. Similarly, a sports implement in flight is often considered a particle for the purpose of mechanical analysis. It must be kept in mind, however, that linear velocities are vector quantities. As such, they must have common units of magnitude and must be acting in the same direction if they are to be combined algebraically. The same comment applies to displacements and accelerations. Therefore, in the planar case, it is customary to express these vectors in x and y components. This particular form of Cartesian reference frame is convenient because the weight vector can be taken parallel to the Y axis. In addition, when air resistance is disregarded, freely falling bodies experience a constant acceleration due to gravity in the vertical or Y direction and a zero acceleration in the horizontal or X direction. After all the necessary calculations have been made using two orthogonal components, a resultant vector indicating both magnitude and direction can be determined.

The anatomical structure of the human body, which may be represented mechanically as a linked system of rigid segments moving about axes of rotation through the joints, lends itself to a consideration of angular position, velocity and acceleration of the limbs. In such an analysis, it is helpful to record the relationship between position and time of several segments on one graph. When similar plots of the angular velocities and angular accelerations are made, the major roles of the limbs in the skill and their interactions may be more clearly appreciated and interpreted. Take for example a

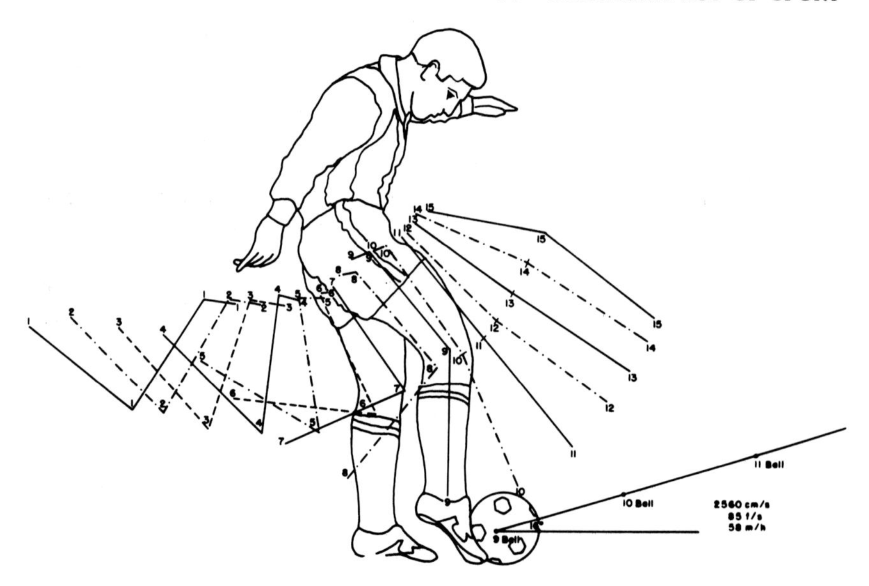

FIGURE 4-1. *Composite Tracing of a Straight Approach Soccer Kick (66 frames/sec) (S. Plagenhoef,* Patterns of Human Motion: A Cinematographic Analysis. *© 1971, p. 99. By permission of Prentice-Hall, Inc., Englewood Cliffs, N.J.).*

soccer kick with a straight approach performed by a skilled player. The successive angular positions of the thigh and shank can be measured from one of Plagenhoef's composite tracings (1971) (Figure 4-1)† and then plotted as functions of time (Figure 4-2). To reduce the effect of measurement and experimental error upon subsequent differentiated functions, the data may be smoothed using either manual or grapho-numerical techniques.‡ From these relationships, the angular velocities of the segments are calculated (Figure 4-3). Smoothed angular velocity-time curves may also provide the basis for studying angular acceleration.

Careful analysis of these curves indicates a great deal about the angular motion of the limbs in the performance. Questions such as the following should be considered. What are the differences in the angular velocity patterns of the segments? Is the distal segment at its maximum angular velocity just prior to contact? Is the distal segment still accelerating at ball contact or is it maintaining a con-

†In this example, a composite film tracing is utilized to illustrate the principles of kinematics. When there is a large amount of data to be analyzed, however, it is more efficient and accurate to use a motion analyzer to measure angles directly from the film or to calculate them from the coordinates of segmental endpoints with the aid of a computer.

‡Information on basic smoothing techniques is included in Appendix C.

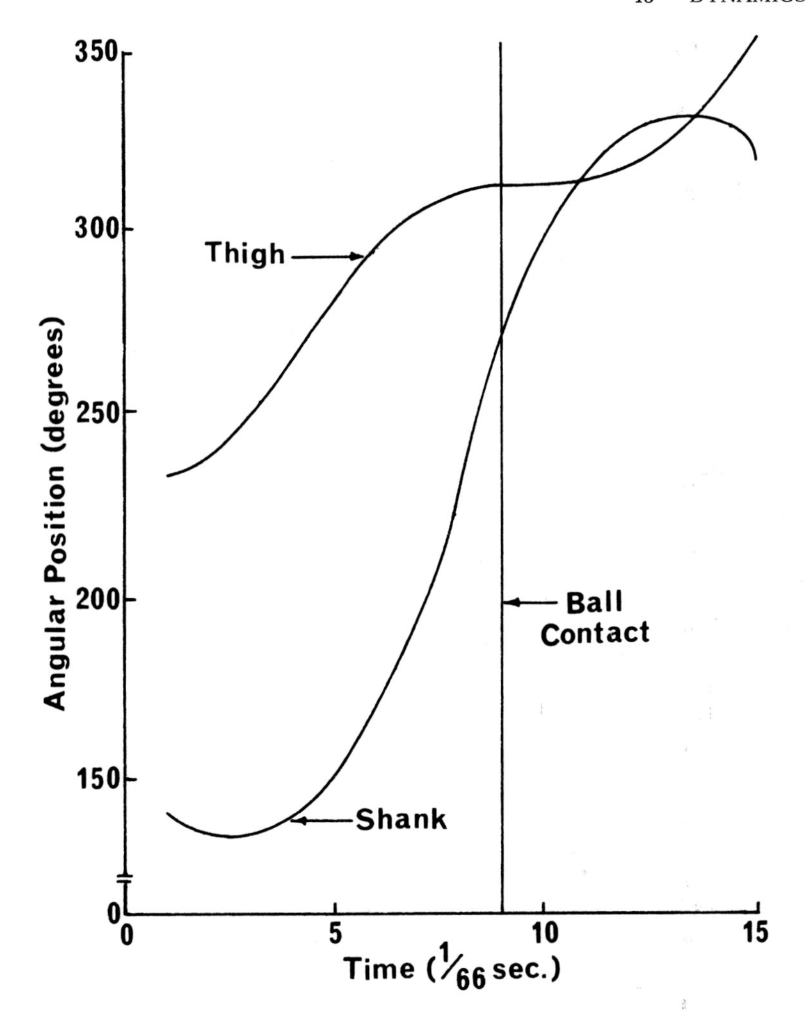

FIGURE 4-2. *Angular Position-Time Relationships of the Thigh and Shank in the Soccer Kick.*

stant velocity? What are the magnitudes of the angular velocities and angular accelerations throughout the range of motion as well as at ball contact? The answers to these and other questions related to the rates of rotation of the body segments, however, do not provide a total description of the mechanics of the skill.

A complete mechanical analysis must incorporate both linear and angular motions of the limbs. In the case of the soccer kick, the absolute linear velocity of the instep at contact with the ball is a function not only of the angular velocity of the thigh and shank but also of the linear velocity of the body as a whole. The same statement

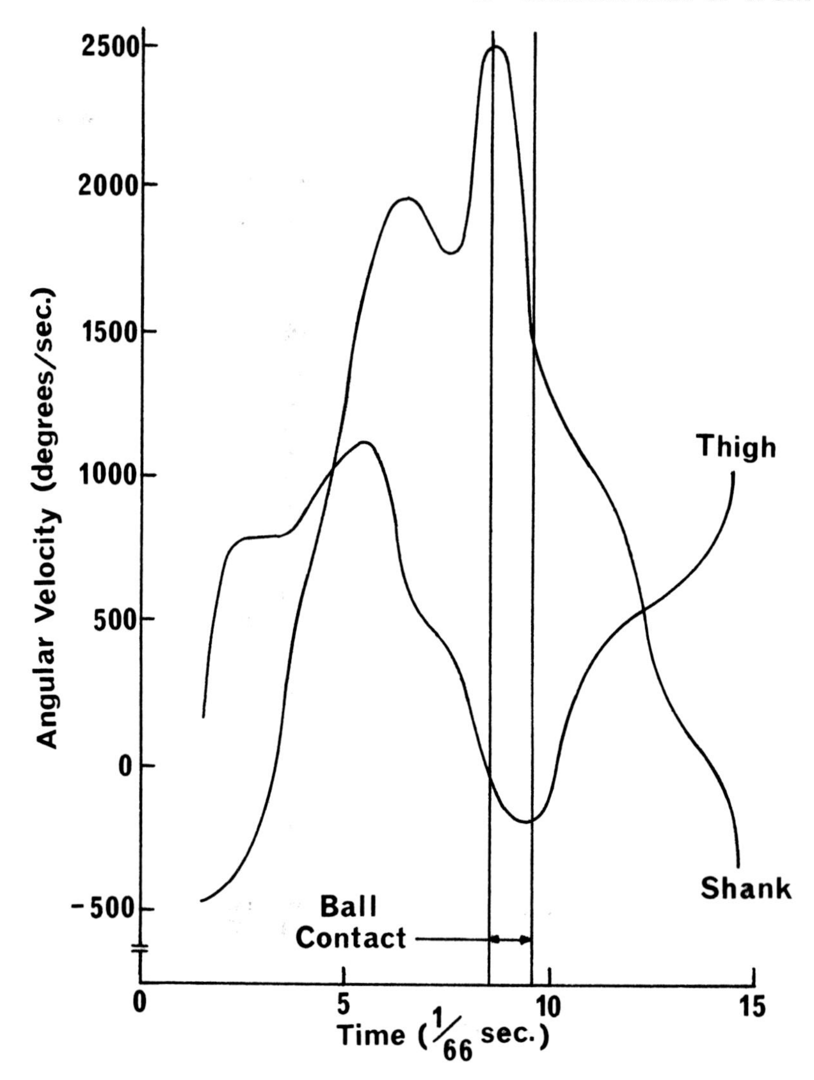

FIGURE 4-3. *Angular Velocity-Time Relationships of the Thigh and Shank in the Soccer Kick.*

can be made concerning displacement and acceleration. It is possible, however, to calculate the linear displacement, velocity and acceleration of any point on a body segment from basic quantitative data describing the mechanics of the performance.

Consider a rigid segment represented by vector **r** with one end designated A and the other B. When this segment moves, the absolute displacement of A is equal to the relative displacement of A with respect to B (assuming for the moment that B is fixed and that the

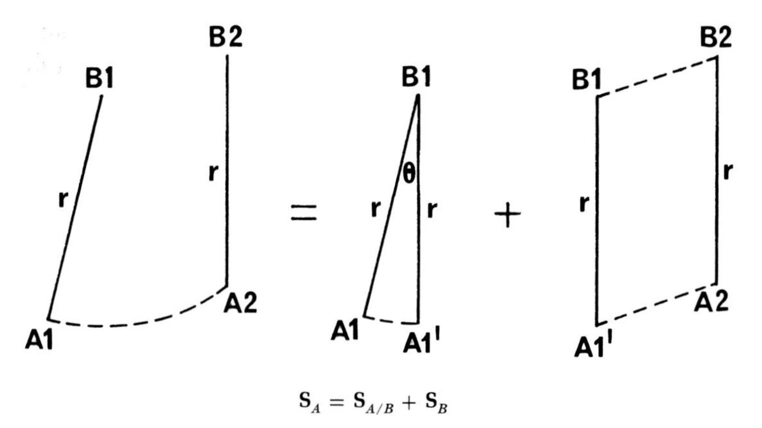

$$\mathbf{S}_A = \mathbf{S}_{A/B} + \mathbf{S}_B$$

FIGURE 4-4. *Displacement of One End of a Segment Experiencing General Plane Motion.*

body is rotating about it) plus the absolute displacement of B. This relationship, shown in Figure 4-4, may be written as follows:

$$\mathbf{s}_A = \mathbf{s}_{A/B} + \mathbf{s}_B$$

in which the magnitude of the relative linear displacement $\mathbf{s}_{A/B} = r * \Delta\theta$. In the case of the linear velocity of A:

$$\mathbf{v}_A = \mathbf{v}_{A/B} + \mathbf{v}_B$$

in which the relative linear velocity $\mathbf{v}_{A/B} = \boldsymbol{\omega} \times \mathbf{r}$ with magnitude $r * \omega$ and direction at right angles to the segment. Similarly, the absolute linear acceleration of A is determined from the relationship:

$$\mathbf{a}_A = \mathbf{a}_{A/B} + \mathbf{a}_B.$$

The relative acceleration can be divided into a normal and a tangential component.

$$\begin{aligned}\mathbf{a}_{A/B} &= (\mathbf{a}_{A/B})_n + (\mathbf{a}_{A/B})_t \\ &= \boldsymbol{\omega} \times (\boldsymbol{\omega} \times \mathbf{r}) + \boldsymbol{\alpha} \times \mathbf{r}.\end{aligned}$$

The normal component has a magnitude $r * \omega^2$ and is directed from A toward B. It is related to the time rate of change of the direction of the velocity. The tangential component reflects the change in the magnitude of the velocity with respect to time. Its magnitude is $r * \alpha$ and its direction is tangent to the path of A.

These relationships can be extended to two or more segments moving in the same plane. The analysis of such a multi-segment motion generally begins at the end of a segment which has a known absolute linear velocity. A foot planted firmly on the ground or a hand maintaining a grip on a fixed support illustrate convenient starting points since they have zero velocity at the contact surface. If the corresponding segment length, position and angular velocity are also known, the absolute linear velocity of the other end of the limb can be determined. With similar information on adjacent segments, the analysis can proceed from one limb to the next.

An approximation of the absolute linear kinematic parameters for any point on a body segment can be obtained by assuming that the particular point moves in a straight line between successive positions. Then the formulae $\mathbf{v} = \frac{\Delta \mathbf{r}}{\Delta t}$ and $\mathbf{a} = \frac{\Delta \mathbf{v}}{\Delta t}$ can be used. Provided that the sampling rate is sufficiently high, the error introduced by this simplification will not be significant.

KINETICS

Kinetics is that section of dynamics concerned with the forces initiating and altering motion. It is the highest level of mechanical analysis and holds the greatest promise for increasing our understanding of the intricacies of human motion in sport. Because of the complexities introduced by the nature of the body, few investigators in physical education have fully explored the potential of kinetic analysis. Traditionally, this aspect of dynamics has been oversimplified. The athlete has been treated as a particle rather than as a linked system. Passing references to the concept of moment of inertia have seldom been translated into quantitative terms. A more rigorous approach must be adopted both in experimental research and theoretical calculations if the contribution of kinetic analysis is to be more fully realized.

Mass and Moment of Inertia. While kinetics is primarily concerned with the influence of forces, it also includes a consideration of the parameters of mass and moment of inertia. These are unchangeable attributes of rigid bodies representing resistance to linear and angular acceleration respectively. Kinetic analysis of the translation of a system requires a knowledge of its mass while both mass and moments of inertia must be known when investigating rotation, general plane motion and general space motion.

In physical terms, moment of inertia is defined as follows:

$$\sum_{i=1}^{n} m_i * r_i^2 \qquad \text{or} \qquad \int r^2\, dm$$

in which m_i and dm represent one of the mass particles and r_i and

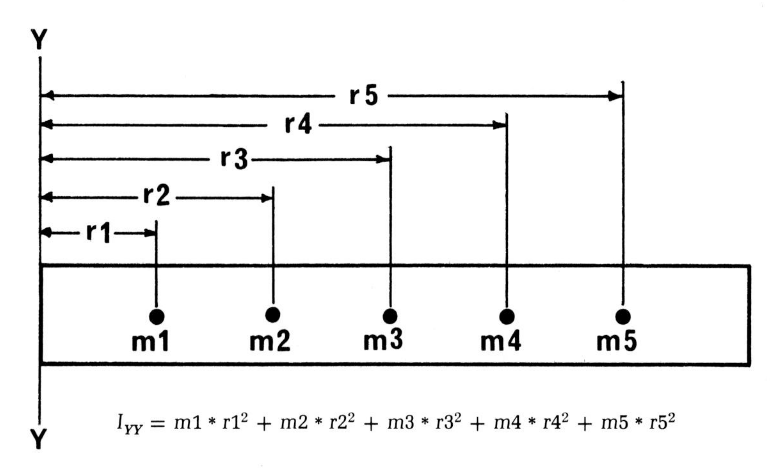

$$I_{YY} = m1 * r1^2 + m2 * r2^2 + m3 * r3^2 + m4 * r4^2 + m5 * r5^2$$

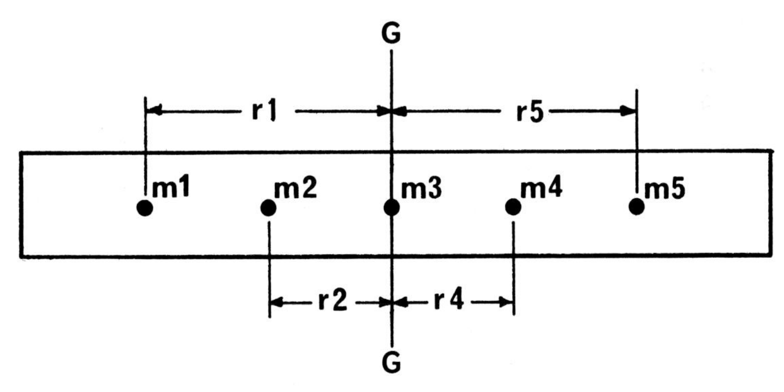

$$I_{GG} = m1 * r1^2 + m2 * r2^2 + m4 * r4^2 + m5 * r5^2$$

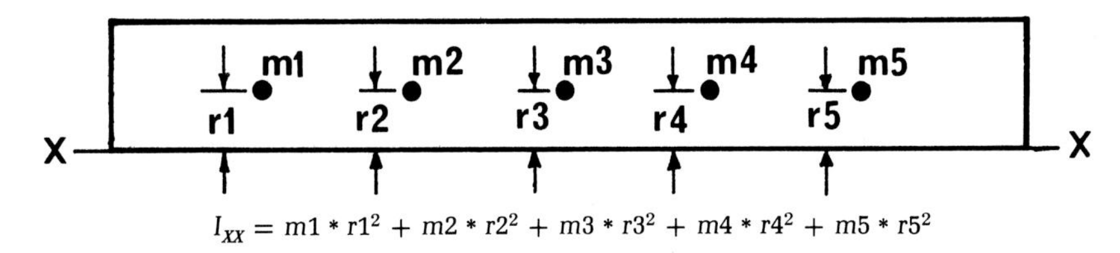

$$I_{XX} = m1 * r1^2 + m2 * r2^2 + m3 * r3^2 + m4 * r4^2 + m5 * r5^2$$

FIGURE 4-5. *Moments of Inertia of an Idealized Body.*

r, the corresponding perpendicular distance of the mass particle from the particular axis of rotation. As can be seen in Figure 4-5, a value denoting moment of inertia has no meaning unless a definite axis is specified. Thus, a rigid body has an infinite number of moments of inertia, one for each axis which can be drawn through an infinite number of points. However, discussion is usually limited to moments of inertia with respect to axes passing through the mass or joint centers. If the body has three axes of symmetry, the moments of inertia about these axes are termed the principal moments of inertia of the body.

The substitution of logical values into the examples given in Figure 4-5 will illustrate the fact that the smallest moment of inertia for any one of a series of parallel axes will be the moment of inertia about an axis through the mass center. It can also be shown that the moment of inertia with respect to any axis parallel to one through the mass center is equal to the moment of inertia about the latter axis plus the product of the mass and the square of the perpendicular distance between the parallel axes (transfer distance) (Figure 4-6). This relationship, known as the parallel axis theorem, is often applied to human body segments.

An expression commonly used in considerations of moments of inertia is radius of gyration (k). It is actually a calculation made after the fact in that the moment of inertia must be known before the radius of gyration can be determined.

$$k = \sqrt{\frac{I}{m}}$$

$$I = mk^2$$

Although it does not exist as a physical entity, k may be defined as "the distance from the axis of rotation to an assumed point where the concentrated total mass of the body would have the same moment of inertia as it does in its original distributed state" (Drillis and Contini, 1966, p. 5). It should be stressed that the radius of gyration *cannot* be used interchangeably with the distance from the axis of rotation to the center of mass of the body. They are two separate entities (Figure 4-7).

General Approach to Kinetic Analysis. The selection of the most appropriate approach to the solution of a particular kinetics problem is an important key to the success of the investigation. Initially, three steps should be followed in an attempt to determine the best method. First, the system under consideration must be carefully defined. Does it encompass the whole body of the athlete? Should an associated sports implement such as a golf club or hockey stick also be included? Is it desirable to limit the analysis to only one or two limbs? Upon this basis, the system should be classified as a particle, rigid (or quasi-rigid) body or a linked system. Such a classification will influence the subsequent selection of the equations of motion.

Secondly, all the external forces acting upon a system must be identified and indicated on a free body diagram. Their points of

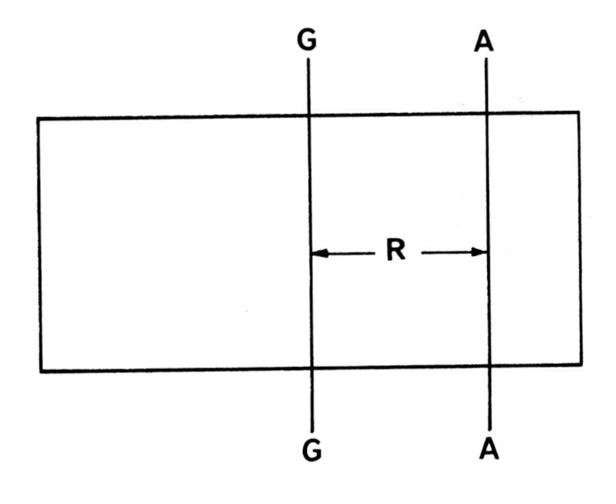

$$I_{AA} = I_{GG} + mR^2$$
$$I_{GG} = I_{AA} - mR^2$$

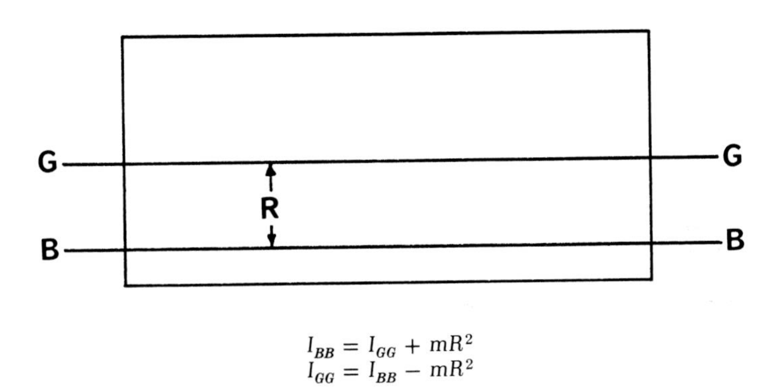

$$I_{BB} = I_{GG} + mR^2$$
$$I_{GG} = I_{BB} - mR^2$$

FIGURE 4-6. *Parallel Axis Theorem.*

application and lines of action should be specified. At this stage, it is advisable to identify both the known and unknown forces and distances in the problem at hand. An excessive number of unknowns may require a redefinition of the system or of the problem.

Finally, consideration must be given to the nature of the motion as the basis for selecting the particular method of analysis. Since many of the external forces fluctuate significantly during the execution of a skill, it may be desirable to examine their magnitudes and directions at particular instants during the course of the motion.

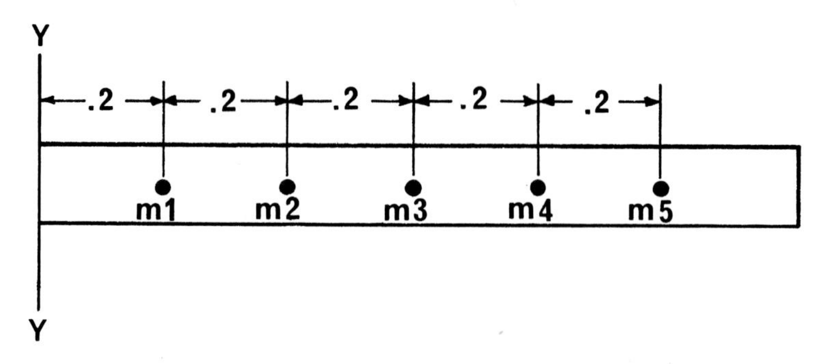

Assuming the idealized body is composed of five .3 slug mass particles located .2 feet apart:

$$
\begin{aligned}
I_{YY} &= \Sigma m_i * r_i^2 \\
&= .3 * .2^2 + .3 * .4^2 + .3 * .6^2 + .3 * .8^2 + .3 * 1.0^2 \\
&= .3 * .04 + .3 * .16 + .3 * .36 + .3 * .64 + .3 * 1.0 \\
&= .012 + .048 + .108 + .192 + .300 \\
&= .660 \text{ slug-feet}^2
\end{aligned}
$$

Since $I_{YY} = mk^2$ and, in this case, $m = 1.5$ slugs

$$k = \sqrt{\frac{I_{YY}}{m}} = \sqrt{\frac{.66}{1.5}} = \sqrt{.44} = .663 \text{ feet}$$

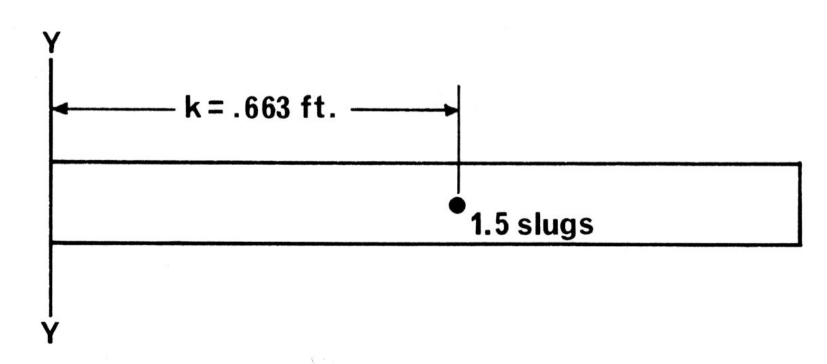

FIGURE 4-7. *Radius of Gyration of an Idealized Body.*

The force-mass-acceleration approach would be suitable for such an instantaneous analysis of the forces concerned. When it is convenient to investigate forces acting over specific distances, then equations of motion derived from work-energy principles may be employed. Impulse-momentum relationships, on the other hand, are suited for the analysis of forces acting during given time periods and must be used in all cases of impact. The application of this general approach to the kinetic analysis of biomechanical problems is illustrated by the following examples.

Ground Reaction Force. In all sports, the athlete is acted upon by a ground reaction force at some time during his performance. The tangential component of this force provides the necessary friction for locomotion while the normal component is of particular importance in obtaining the height required for a basketball rebound, volleyball block, high jump or dance leap. Although a force platform† is the best type of instrumentation for investigating the fluctuations of this force, it is also possible to use a good quality weigh scale to estimate its vertical component during rather restricted movements such as crouching or raising the arms. Figure 4-8 illustrates a laboratory exercise in which a scale, Polaroid sequence camera, timing display and reference distance are employed to record the displacement of the subject's mass center (approximated by a point on the hip) and the vertical portion of the ground reaction force. While such an exercise serves to introduce the basic concepts of reaction force variations, the weigh scale apparatus is not suitable for studying motion of an explosive nature.

†Refer to Chapter 7, pages 176–177.

FIGURE 4-8. *Laboratory Procedure for Studying Fluctuations in the Vertical Component of the Ground Reaction Force.*

The preparatory movements for a standing vertical jump exemplify the general pattern of the ground reaction generated in response to the actions of the trunk and limbs. For a kinetic analysis of this motion, the system is defined as the body of the jumper. In this particular instance, rotation can be neglected without introducing significant error into the calculations. Therefore, the system may be represented as a particle (the mass center) with mass equal to that of the subject. The free body diagram (Figure 4-9) indicates that the body weight acts vertically downward from the mass center. The force platform upon which the jumper is standing exerts a reaction force which can be expressed as a vertical (normal) and a horizontal (tangential) component.

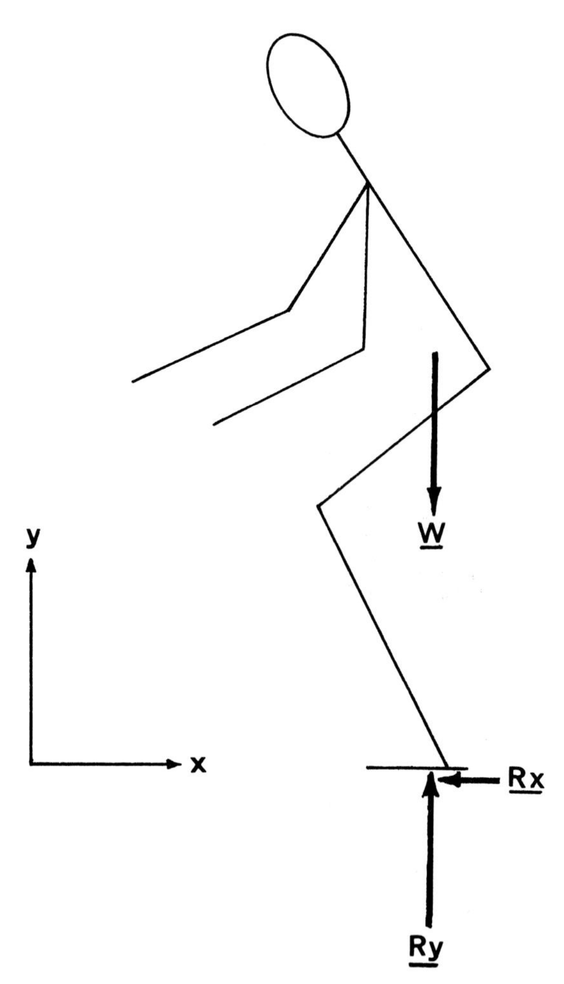

FIGURE 4-9. *Free Body Diagram of a Vertical Jump Take-off.*

An examination of the magnitudes of the forces at successive points during the take-off is best accomplished by a direct application of Newton's second law of motion (force-mass-acceleration method). Since the forces are restricted to a single plane, the equations of motion may be expressed as:

$$\Sigma F_x = ma_x \qquad \Sigma F_y = ma_y$$
$$R_x = ma_x \qquad R_y - W = ma_y$$

in which x and y represent their respective coordinate directions; a is the acceleration of the mass center of the jumper; m is the body mass, which remains constant; and R_x and R_y indicate the horizontal and vertical components of the reaction force.

Since height is the prime concern in this example, detailed consideration will be restricted to the vertical portion of the take-off. The equation of vertical motion can be rearranged to isolate R_y, the vertical reaction force component:

$$\Sigma F_y = ma_y$$
$$R_y - W = ma_y$$
$$R_y = W + ma_y.$$

Careful examination of this relationship reveals that the magnitude of R_y will be less than body weight if the jumper's resultant acceleration is negative and greater than body weight if a_y is positive. In the case of zero acceleration, the vertical reaction force will equal the weight. Thus, the force-mass-acceleration approach defines the motion of the mass center at specific points in time. For example, when a vertical reaction force of 234 pounds is registered for a 144-lb jumper, the vertical acceleration of his mass center at that instant is 20 ft/sec/sec.

$$R_y = W + ma_y$$
$$R_y = W + \frac{W}{g} a_y$$
$$234 = 144 + \frac{144}{32} a_y$$
$$a_y = 20 \text{ ft/sec/sec}$$

Table 4-3 and Figure 4-10 provide information on the kinematics and kinetics of the vertical jump take-off. The calculations of the velocity and acceleration of the mass center of a 144-lb jumper and the associated vertical ground reaction force modelled after experimental data (Gerrish, 1934) furnish several insights into the skill. Initially, in preparation for the jump, the body accelerates downward in what is considered a negative direction. As a result, the vertical

Table 4-3. Kinematics and Kinetics of a Vertical Jump Take-off†

Time (sec)	*Position (ft)*	*Displacement (ft)*	*Velocity (ft/sec)*	*Change in Velocity (ft/sec)*	*Acceleration (ft/sec²)*	*Reaction Force (lb)*	*Comments*
0.00	3.40						
		−0.01	−0.20				
0.05	3.39			−0.20	−4.00	126	
		−0.02	−0.40				Increasing
0.10	3.37			−0.20	−4.00	126	
		−0.03	−0.60				
0.15	3.34			−0.40	−8.00	108	
		−0.05	−1.00				Negative
0.20	3.29			−0.60	−12.00	90	Acceleration
		−0.08	−1.60				
0.25	3.21			−1.20	−24.00	36	
		−0.14	−2.80				
0.30	3.07			−0.60	−12.00	90	Decreasing
		−0.17	−3.40				
0.35	2.90			−0.40	−8.00	108	
		−0.19	−3.80				Lowest
0.40	2.71			0.60	12.00	198	Velocity
		−0.16	−3.20				
0.45	3.55			0.80	16.00	216	
		−0.12	−2.40				Increasing
0.50	2.43			1.00	20.00	234	
		−0.07	−1.40				
0.55	2.36			1.40	28.00	270	
		0.00	0.00				
0.60	2.36			1.80	36.00	306	Positive
		0.09	1.80				Acceleration
0.65	2.45			2.60	52.00	378	
		0.22	4.40				
0.70	2.67			2.80	56.00	396	
		0.36	7.20				
0.75	3.03			3.20	64.00	432	
		0.52	10.40				Decreasing
0.80	3.55			1.60	32.00	288	
		0.60	12.00				
0.85	4.15			−0.60	−12.00	90	
		0.57	11.40				Take-off
0.90	4.72			−1.60	−32.00	0	
		0.49	9.80				Negative
0.95	5.21			−1.60	−32.00	0	Acceleration
		0.41	8.20				
1.00	5.62						

†Refers to the vertical movement of the mass center of a 144-lb jumper and the vertical component of the reaction force. The acceleration due to gravity is assumed to be −32.00 ft/sec/sec.

reaction force drops below body weight. While still moving downward, as indicated by the position values and negative velocities, the body begins to accelerate positively. This positive acceleration is reflected in a vertical reaction force which exceeds the weight of the subject. If the measurements and calculations had been made

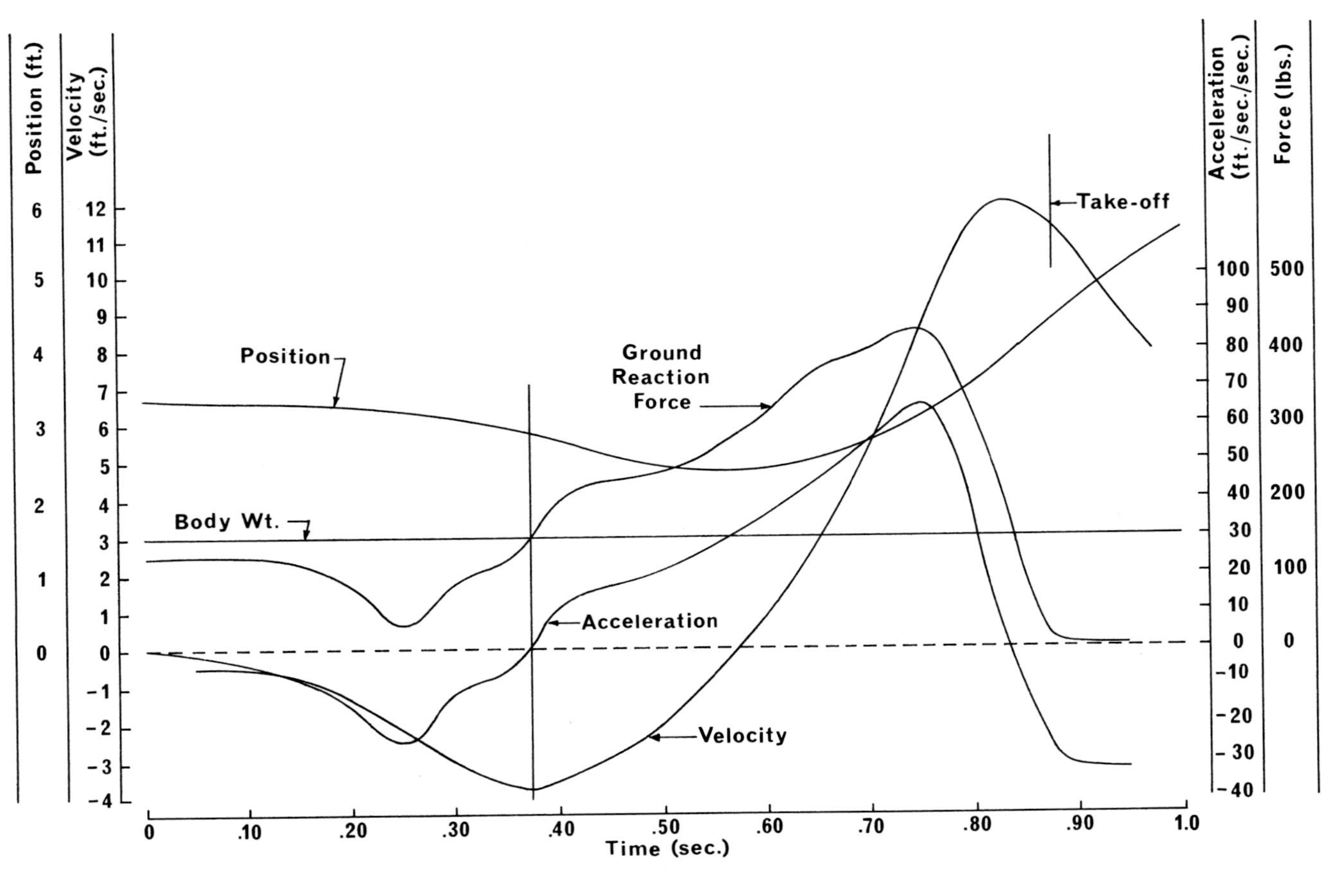

FIGURE 4-10. *Vertical Kinematics and Kinetics of the Vertical Jump Take-off.*

at more frequent time intervals, it would be noted that the maximum negative velocity coincides with a vertical reaction force equivalent to body weight. The velocity then increases to a near maximum positive value prior to take-off. The body, however, begins to decelerate just before leaving the ground as a result of decreasing limb velocities. Once the jumper is free in the air, he is subject to the influence of only one force, the weight of his body. Consequently, R_y is zero, his mass center maintains a constant negative acceleration due to gravity and his velocity continues to decrease.

The acceleration of the mass center of the jumper is the resultant of the accelerations of all of his individual segments. Thus,

$$m\mathbf{a} = m_1\mathbf{a}_1 + m_2\mathbf{a}_2 + m_3\mathbf{a}_3 + \cdots + m_n\mathbf{a}_n$$

$$\mathbf{a} = \frac{m_1\mathbf{a}_1}{m} + \frac{m_2\mathbf{a}_2}{m} + \frac{m_3\mathbf{a}_3}{m} + \cdots + \frac{m_n\mathbf{a}_n}{m}$$

in which $m_1 \ldots m_n$ represent the masses of the segments and $\mathbf{a}_1 \ldots \mathbf{a}_n$ indicate the accelerations of their respective mass centers. From this relationship, it is evident that the contribution of each segment to the total acceleration is proportional to both its mass and the acceleration of its segmental mass center. To understand the interrelationships and influences of various parts of the body during the preparatory phases of a jump, the acceleration of each segment should be considered separately and then related to the whole movement.

In addition to examining the values of the forces and the resultant motion of the jumper at successive points in time, it is important to investigate the change in his motion which occurs during the take-off. When a force acts over a time interval (impulse), it causes a change in the "quantity of motion" known as momentum. In this particular instance in which the jumper is treated as a mass point, the analysis may be restricted to a consideration of linear impulse and linear momentum. The linear impulse-momentum relationship is derived from Newton's second law of motion:

$$\Sigma\mathbf{F} = m\mathbf{a}$$

$$\Sigma\mathbf{F} = \frac{m\,d\mathbf{v}}{dt}$$

$$\int \mathbf{F}\,dt = \int \frac{m\,d\mathbf{v}\,dt}{dt}$$

$$\int_{t_i}^{t_f} \mathbf{F}\,dt = \int_{t_i}^{t_f} \frac{m\,d\mathbf{v}\,dt}{dt}$$

in which i and f represent the initial and final values of the integration interval. Since the mass remains constant,

$$\int_{t_i}^{t_f} \mathbf{F}\, dt = m \int_{t_i}^{t_f} \frac{d\mathbf{v}\, dt}{dt}$$
$$= m(\mathbf{v_f} - \mathbf{v_i})$$
$$= \Delta m\mathbf{v}.$$

Therefore, linear impulse $\left(\int_{t_i}^{t_f} \mathbf{F}\, dt\right)$ is always equal to the change in linear momentum ($\Delta m\mathbf{v}$).

The designation **F** includes all the forces acting upon the system. They can be resolved along mutually perpendicular axes in order to examine the momentum changes in specific directions. In the present example, the horizontal, and more particularly, the vertical directions are of concern. Thus,

$$\Sigma F_x = R_x$$
$$\int_{t_i}^{t_f} R_x\, dt = m(v_{x_f} - v_{x_i})$$
$$\Sigma F_y = R_y - W$$
$$\int_{t_i}^{t_f} (R_y - W)\, dt = m(v_{y_f} - v_{y_i})$$
$$\int_{t_i}^{t_f} R_y\, dt - \int_{t_i}^{t_f} W\, dt = m(v_{y_f} - v_{y_i}).$$

The impulse of each force is equal to the area under its force-time curve.

Figure 4-11 shows the vertical forces and force components acting upon the jumper during the .865 second required for the take-off. Since the weight remains constant, the area beneath its "curve" is a rectangle with base .865 second and height 144 pounds yielding an impulse of 124.56 lb-sec. The vertical reaction force, however, fluctuates in response to the acceleration of the individual body segments. The total area under its curve may be determined by using a planimeter. Alternately, it can be subdivided into small squares or rectangles of convenient size.† These small areas are then summed to give an estimate of the impulse. Beneath the R_y force-time curve in the present example, there are approximately 143 rectangles (base .05 sec and height 25 lb) each of which represents 1.25 lb-sec. This indicates that the impulse of R_y is 178.75 lb-sec during the specified time interval. Substituting these values into the linear impulse-momentum relationship yields the vertical velocity of the jumper's mass center at take-off.

†Usually graph paper is used for this purpose.

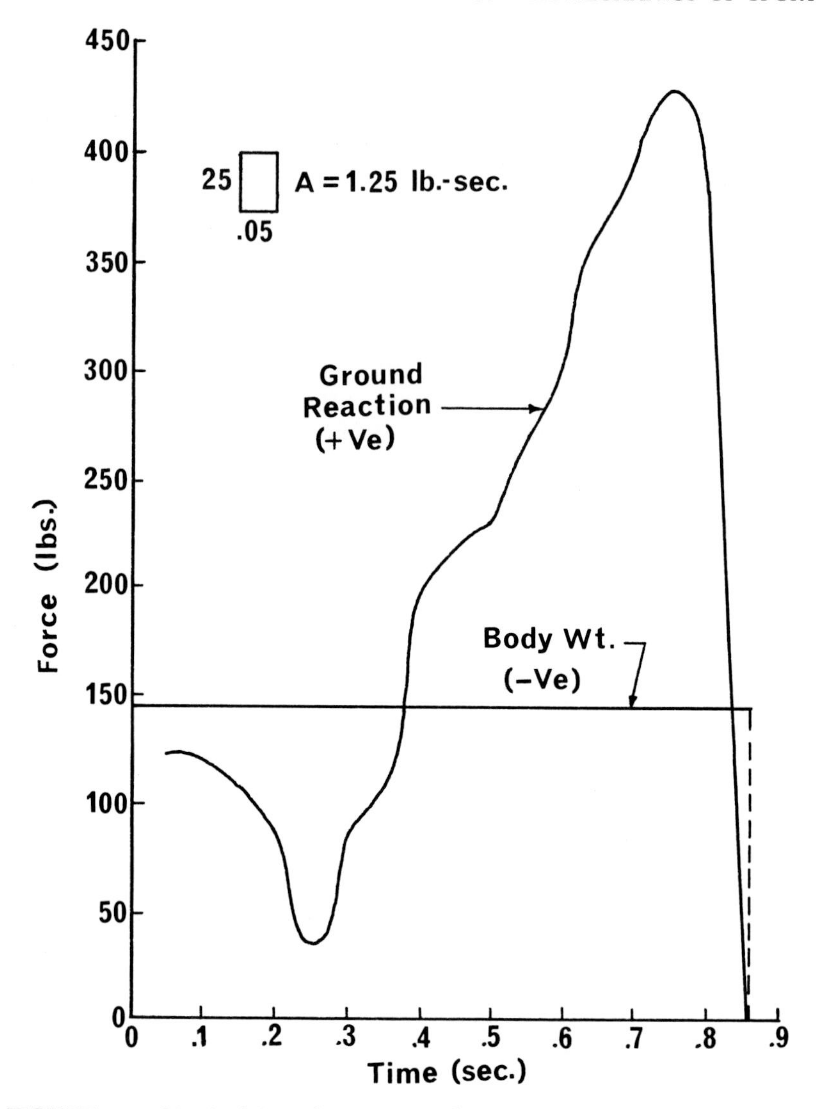

FIGURE 4-11. *Vertical Impulses During the Take-off.*

$$\int_0^{.865} R_y\, dt - \int_0^{.865} W\, dt = m(v_{y_f} - v_{y_i})$$

$$178.75 - 124.56 = \frac{144}{32}(v_{y_f} - 0)$$

$$v_{y_f} = 12.04 \text{ ft/sec}$$

This is in good agreement with the experimental value calculated from displacements of the mass center.

To achieve an effective take-off velocity, a jumper must be able to generate a substantial impulse from the ground reaction force. This force varies directly with the accelerations of the individual segments. Therefore, to develop a large positive vertical reaction component, there must be correspondingly large upward accelerations of the trunk and limbs. Ramey (1972), in studying the long jump take-off, showed that the rapid straightening of the thrust leg provided the major contribution to the vertical impulse. Theoretically, an increase in the impulse of the reaction force can be accomplished either by increasing the magnitude of the force or the time over which it acts. If the time interval is reduced, the reaction force must exhibit a more rapid build up early in the take-off in order to attain an equivalent area beneath the force-time curve.

Muscle Torque. Insight into the action of muscles which are responsible for initiating and controlling the majority of the movements of the body segments may be gained by kinetic analysis. For most practical purposes, however, it is impossible to identify the magnitude and direction of separate muscle forces in a dynamic situation. Because of individual differences in anatomical structure, the exact location of the bony attachments of muscles cannot be generalized. In addition, if every muscle were treated separately in the mechanical analysis, the number of unknowns would greatly exceed the equations of motion, thus making the problem indeterminate. For these reasons, it is customary to speak of idealized resultant muscle forces (Figure 4-12) and to study the resultant muscular torques which they generate. The force-mass-acceleration method of deriving the equations of motion for rigid bodies may be used to determine the values of these torques at successive instants in the movement. Subsequently, the pattern of their variations as functions of time may be investigated in relation to the kinematics of the motion.

Dillman (1970) employed these principles in studying the kinetics of the recovery leg in sprint running. He represented the lower limb as a system of three homogeneous rigid bodies and considered each individually in a sequential analysis which began with the foot and then progressed to the lower leg and thigh. The free body diagrams of the segments (Figure 4-13 through Figure 4-17) indicate their segmental weights directed vertically downward from their respective mass centers, joint reaction forces represented by horizontal and vertical components, and a resultant muscle force crossing each joint. The latter is the vector sum of all the individual muscle forces, both agonists and antagonists, which act upon a particular joint. The resultant muscle force will cause a torque equivalent to the moments of all the original forces composing it.

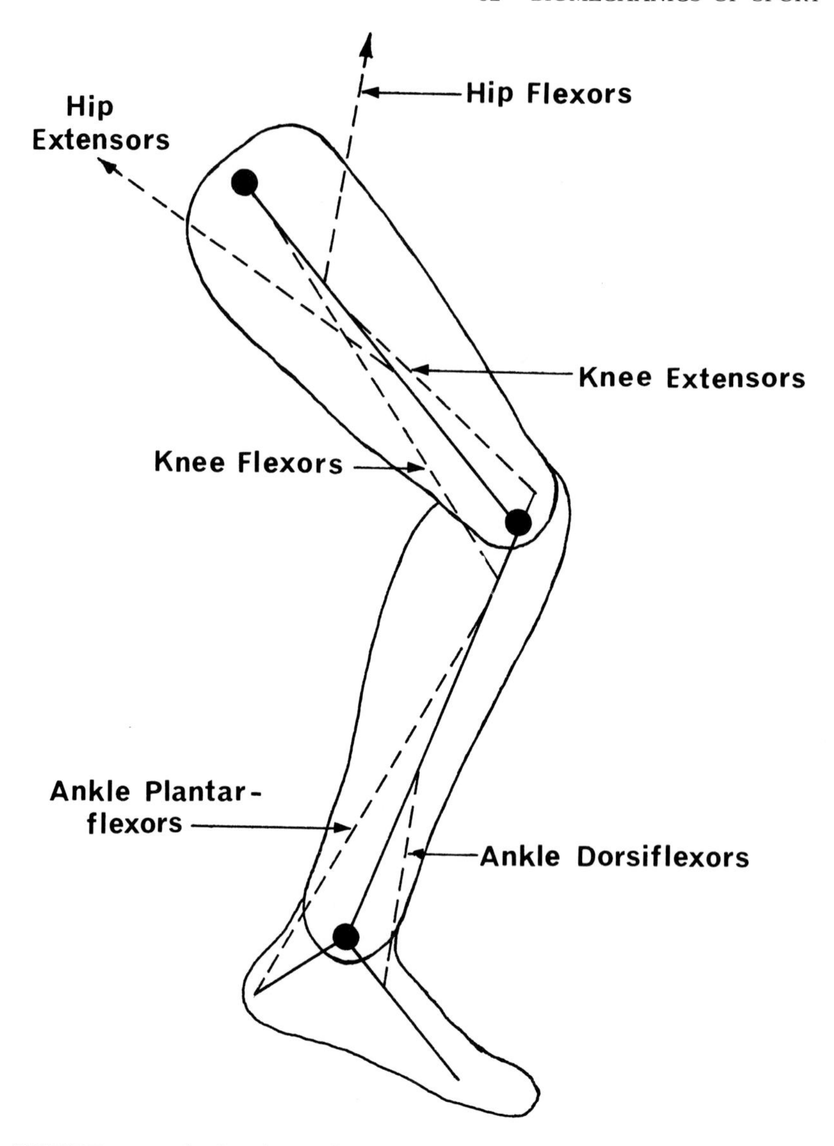

FIGURE 4-12. *Idealized Resultant Muscle Forces of the Lower Limb.*

To achieve consistency in the free body diagrams and to facilitate subsequent computer processing and data analysis, the following conventions have been adopted:

1. Lines of action of forces directed upward and to the right are positive while those directed downward or to the left are negative.
2. Rotation in a counterclockwise direction is considered positive and clockwise rotation negative.

3. All angles are expressed with respect to the right horizontal axis being zero.

4. Resultant force components and the resultant muscular torque at the proximal end of a segment are indicated as positive on the free body diagram while those at the distal end are negative.

5. The symbols employed include:

$W(i)$	segment weight
$JX(i), JY(i)$	x and y components of joint reaction force
$F(i)$	resultant muscle force

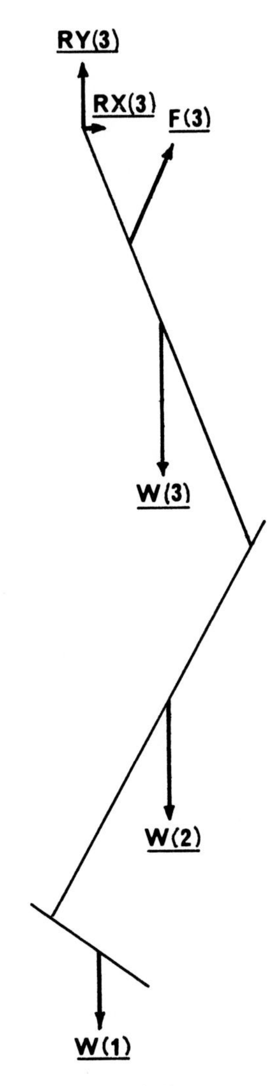

FIGURE 4-13. *Free Body Diagram of the Lower Limb.*

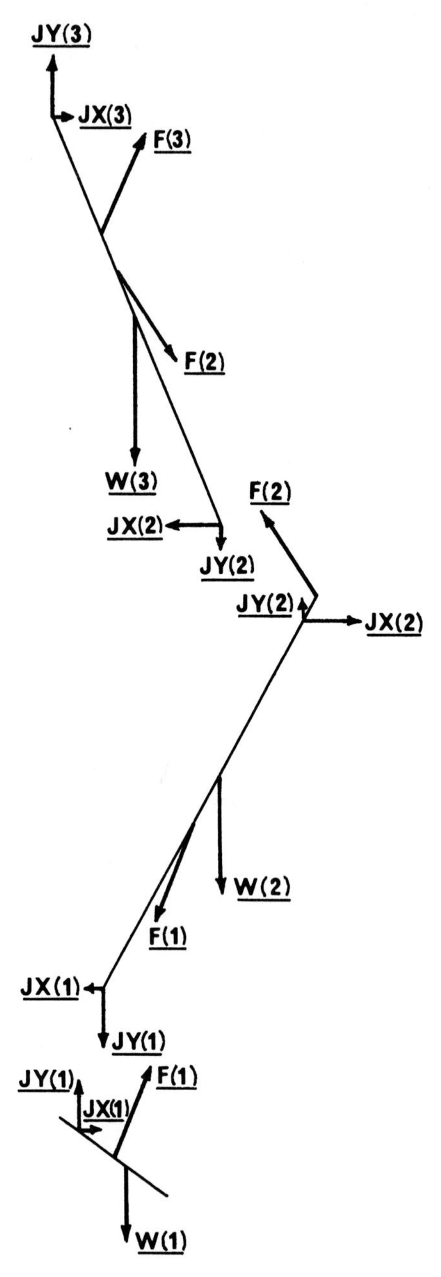

FIGURE 4-14. *Free Body Diagrams of the Three Limb Segments.*

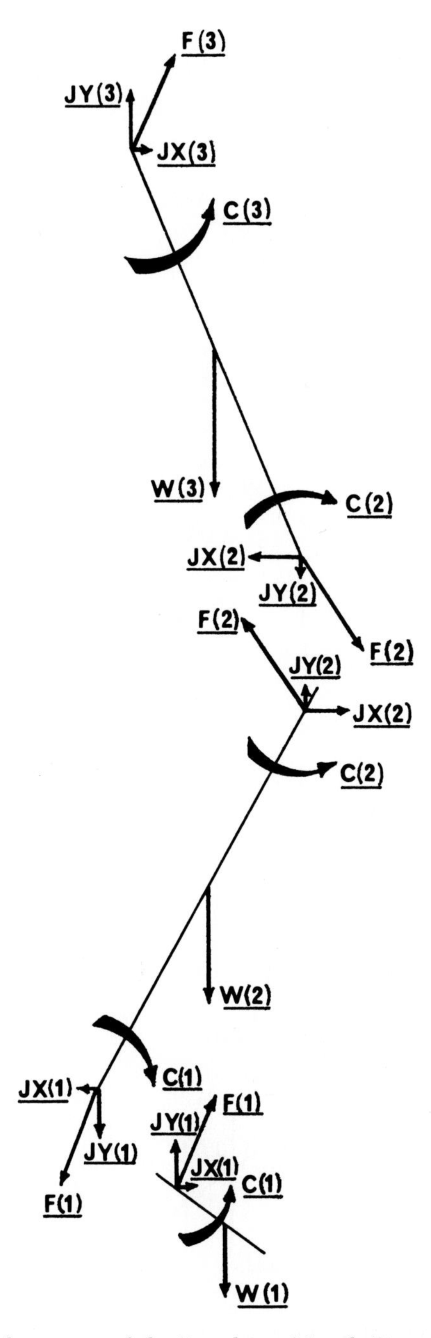

FIGURE 4-15. *Replacement of the Resultant Muscle Force with an Equivalent Force and Couple.*

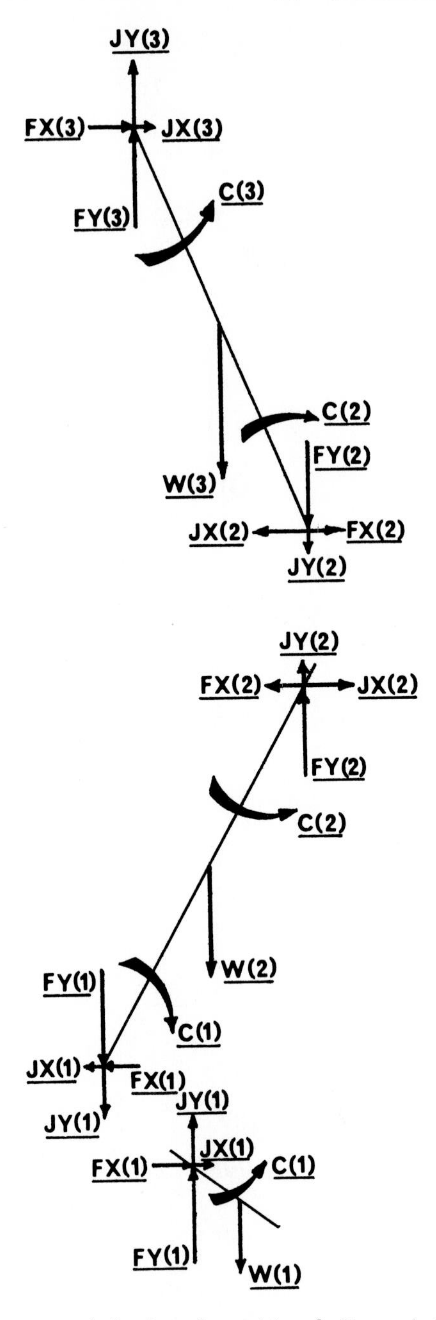

FIGURE 4-16. *Division of the Resultant Muscle Force into Horizontal and Vertical Components.*

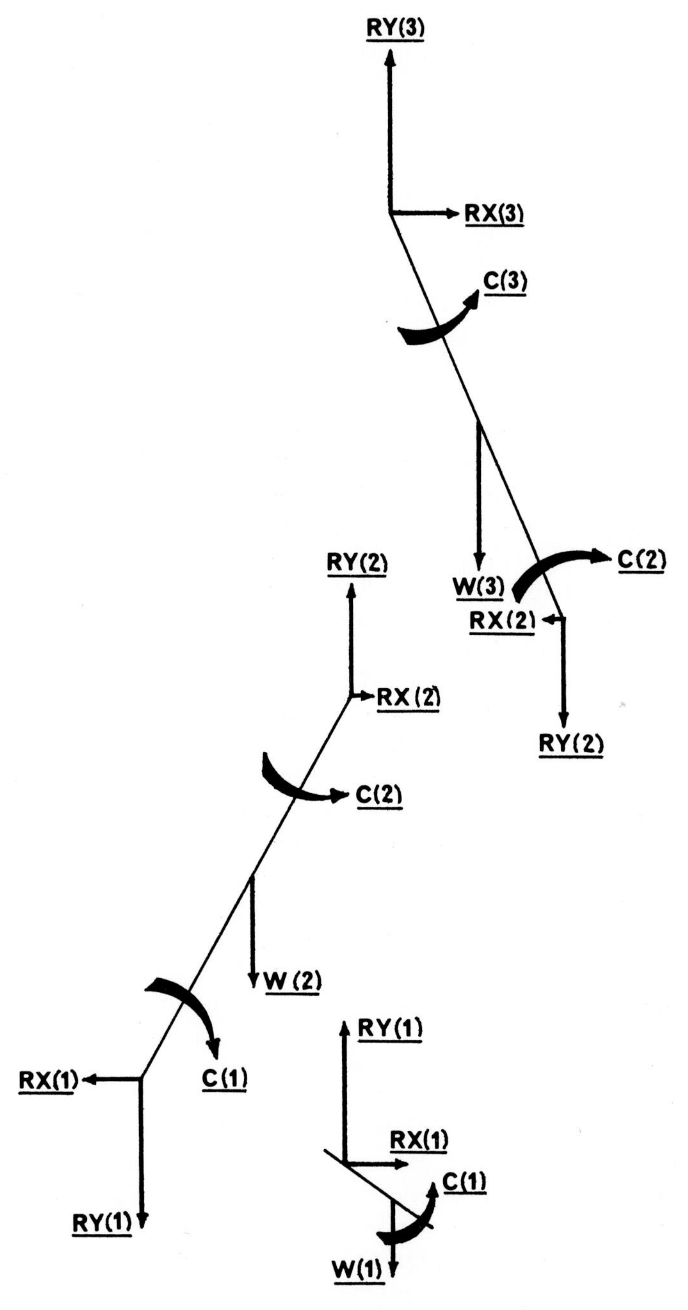

FIGURE 4-17. *Combination of Joint Reaction and Muscle Force Components into Resultant Force Components at the Joint.*

$FX(i)$, $FY(i)$	x and y components of resultant muscle force
$C(i)$	force couple generated by resultant muscle force
$RX(i)$, $RY(i)$	x and y components of the combined resultant muscle and joint reaction forces

in which $i = 1, 3$ to relate the variable designations to particular segments or joints with (1) indicating the foot or ankle; (2), the lower leg or knee; and (3), the thigh or hip joint.

Since the path of the recovery leg is assumed to be restricted to a single plane, three equations of motion can be derived by summing the forces in two orthogonal directions and by summing the moments of force about a given point. Many of these variables in the mechanical analysis including segmental weights, moments of inertia, mass center locations, limb lengths and linear and angular accelerations can either be determined cinematographically or estimated from previous studies. Five unknowns, however, remain in the free body diagram of the foot: the magnitude and direction of the joint reaction force, and the magnitude, direction and point of application of the resultant muscle force. Mathematical manipulations can be employed to combine some of these unknowns into a more manageable number. The resultant muscle force can be replaced by a force of equal magnitude and direction at the joint center and a couple† equivalent to the rotary effect of the original forces about the joint (Figure 4-15). For convenience, the translated muscle force may be indicated by horizontal and vertical components of suitable magnitudes (Figure 4-16). Further, the muscle force components may be combined with those of the joint reaction force (Figure 4-17). Thus, the number of unknowns in the adjusted free body diagram can be reduced to three: the resultant muscular torque, and the magnitude and direction of the combined muscle and reaction force at the joint. Similar adjustments can be made in the free body diagrams of the lower leg and thigh.

In applying the force-mass-acceleration method to derive the equations of motion, the free body diagram of the system is "equated" to a mass-acceleration diagram (Figure 4-18) since both have the same resultant (Meriam, 1966). The mass-acceleration diagram shows the linear acceleration of the mass center (usually in component form) multiplied by the segmental mass as well as a couple equal to the product of the moment of inertia of the segment about an axis through its mass center perpendicular to the plane of motion and the angular acceleration of the segment. While the couple

†The terms couple, torque and moment are used interchangeably to designate the rotational effect of a force with respect to a given point or axis.

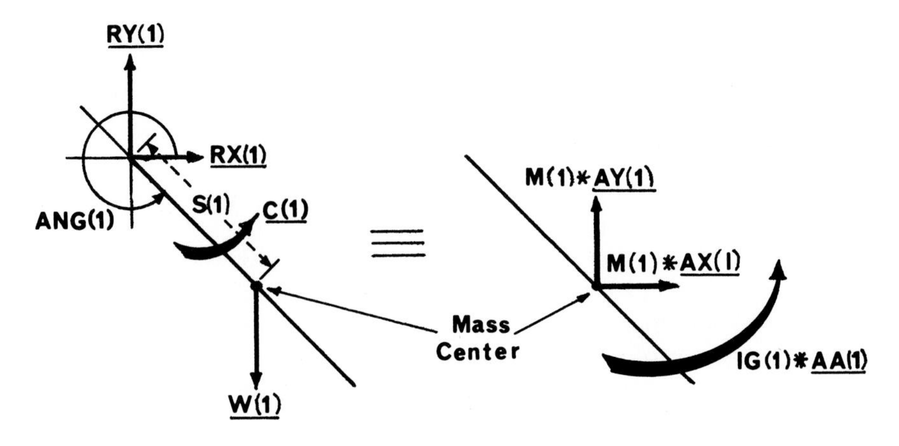

Free Body Diagram Mass-Acceleration Diagram

FIGURE 4-18. *Basis for Deriving Equations of Motion of the Foot.*

can act anywhere in the plane of motion, it is customarily drawn about the mass center.

The three equations of motion of the foot are then obtained by the following:

1. Summing the forces in the horizontal direction and setting them equal to the segmental mass times the horizontal acceleration of the mass center

$$RX(1) = M(1) * AX(1)$$

2. Summing the forces in the vertical direction and setting them equal to the segmental mass times the vertical acceleration of the mass center

$$RY(1) - W(1) = M(1) * AY(1)$$

3. Summing the moments of force about the mass center and setting them equal to the product of the moment of inertia of the segment with respect to its mass center and the angular acceleration of the segment

$$C(1) + RX(1) * S(1) * SIN(ANG(1)) - RY(1) * S(1) * COS(ANG(1)) = IG(1) * AA(1)$$

in which:

$RX(1)$ and $RY(1)$ are the components of the resultant muscle and reaction forces at the ankle joint;

$M(1)$ and $W(1)$ are the mass and weight of the foot respectively;

AX(1) and *AY*(1) are the components of linear acceleration of the mass center of the foot;
C(1) is the resultant muscular torque or couple at the ankle;
S(1) is the distance from the proximal joint to the mass center of the foot;
AA(1) is the angular acceleration of the foot, often designated by the Greek letter alpha (α);
ANG(1) is the angle of the foot with the right horizontal in a counterclockwise direction measured at the ankle joint; and
IG(1) is the moment of inertia of the foot about its mass center or center of gravity.

While the moment equation may sometimes be expressed as

$$\Sigma M_o = I_o \alpha$$

this relationship actually applies only to the special cases in which *o* is a fixed point or the center of mass of the rigid body. A more general approach which does not require these specific conditions may be used. Moments can be summed about any point on the free body diagram and set equal to the sum of the moments about the same point on the mass-acceleration diagram. Forces and mass-acceleration terms whose lines of action do not pass through the chosen point have a rotational effect or moment about that point. For example, moments could be summed conveniently about the ankle joint, yielding the following equation:

$$C(1) - W(1) * S(1) * COS(ANG(1)) = IG(1) * AA(1) + M(1) * AY(1) * S(1) * COS(ANG(1)) - M(1) * AX(1) * S(1) * SIN(ANG(1))$$

which is equivalent to the moment equation derived earlier.

A careful examination of Figure 4-17 reveals that the reaction force, resultant muscle force and/or resultant muscular torque on either side of the joint are equal in magnitude and opposite in direction. If the foot, lower leg and thigh were to be joined together at any stage of the analysis, the muscle and joint forces at the knee and ankle would cancel one another and would not appear explicitly in the equations. Therefore, the values of *C*(1), *RX*(1) and *RY*(1) determined from the equations of motion for the foot can be substituted into the force-mass-acceleration relationships for the lower leg. Similarly, *C*(2), *RX*(2) and *RY*(2) from the lower leg can be used in the equations of motion for the thigh. It should be realized, however, that such a derivation of the equations of motion assumes that the muscles cross only one joint or that the discrepancies introduced by two joint muscles can be considered negligible in the analysis.

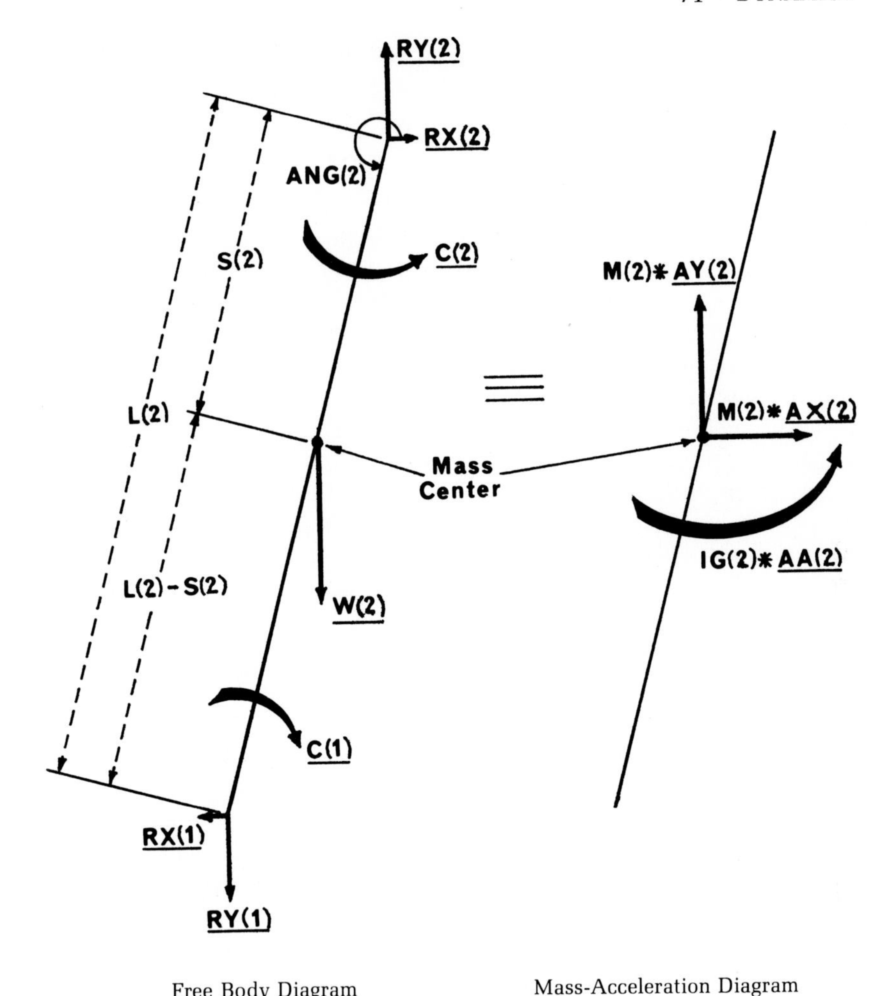

Free Body Diagram Mass-Acceleration Diagram

FIGURE 4-19. *Basis for Deriving the Equations of Motion of the Lower Leg.*

In deriving the equations of motion for the lower leg and thigh, the same principles and variable designations are employed with the addition of L for segment length. The three equations for the lower leg are obtained by the following (Figure 4-19):

1. Summing the forces in the horizontal direction

$$RX(2) - RX(1) = M(2) * AX(2)$$

2. Summing the forces in the vertical direction

$$- RY(1) + RY(2) - W(2) = M(2) * AY(2)$$

3. Summing the moments of force about the mass center of the segment

$$- C(1) + C(2) + RX(1) * (L(2) - S(2)) * SIN(ANG(2)) - RY(1) * (L(2) - S(2)) * COS(ANG(2)) - RY(2) * S(2) * COS(ANG(2)) + RX(2) * S(2) * SIN(ANG(2)) = IG(2) * AA(2)$$

The three equations of motion for the thigh can be developed in a similar manner. When solving these equations for the resultant force components $RX(i)$ and $RY(i)$ and resultant muscle couple $C(i)$, a positive sign indicates that the particular vector is directed in the sense shown on the free body diagram while a negative sign means that the vector acts in the opposite direction.

Utilizing these force-mass-acceleration principles, Dillman (1970) determined the values of the resultant muscular torques at the ankle, knee and hip joints. With these data and a knowledge of the kinematics of the lower limb, he was able to identify the pattern of dominant muscle group activity corresponding to the motion of the recovery leg in sprint running. Other investigators including Elftman (1940), Pearson and his associates (1963) and Plagenhoef (1966, 1971) have also employed this type of approach for studying the dynamics of human performance.

Free Fall Conditions. In sport, it is not uncommon for a performer or his sports implement to experience a condition of non-support or free fall. Some of these activities, which include golf, tennis, badminton, sky diving and ski jumping, are significantly influenced by air resistance. In others, such as those listed in Table 4-4, the effect of air resistance can be disregarded, thereby considerably simplifying the mechanical analysis. Further discussion in this section will be confined to the free fall portion of sports skills in which negligible air resistance can be assumed.

The performance of a skill in an unsupported situation may be classified as general plane motion or general space motion. Thus, the action can be analyzed in terms of the translation of the athlete's mass center and the rotation of the body and its segments. If the movement is of a spatial nature, three components of translation may be considered; one in the vertical direction and two in a horizontal plane at right angles to one another. Likewise rotation is expressed about a vertical or longitudinal body axis as well as about two horizontal axes which usually correspond to the frontal (lateral) and sagittal (anteroposterior) planes of the body. In sport, rotations about the first two are customarily referred to as twisting and somersaulting (Figure 4-20). In aerospace terminology, the three axes are designated yaw, pitch and roll respectively. Although the principal object of

Table 4-4. Sports Skill Objectives†

	Translation		*Rotation*	
	Height	*Horizontal Distance*	*Somersault*	*Twist*
Basketball				
Jump Shot	***			
Rebound	***			
Pass Interception	***	*		
Dance				
Leap	**	**		
Field Events				
High Jump	***	*	*	*
Long Jump	*	***		
Triple Jump	*	***		
Pole Vault	***		*	*
Shot Put	**	***		
Hammer Throw	**	***		
Figure Skating				
Jumps	***	*		***
Gymnastics/Tumbling				
Somersaults	**	*	***	
Twisting Somersaults	**	*	**	***
Leaps	***	**		
Vaults	**	**	**	*
Springboard Diving				
Nontwisting Dives	***	*	*	
Somersaults	**	*	***	
Twisting Somersaults	**	*	**	***
Trampoline				
Bouncing	***			
Somersaults	**		***	
Twisting Somersaults	**		**	***
Volleyball				
Block	***			
Spike	***	*		

†Asterisks indicate degree of importance of objectives ranging from least (*) to most important (***).

a skill may be to achieve the maximum value of only one of these six components such as height in the jump shot, it is more common to seek an optimal combination of two or more as in the case of the twisting somersault.

The trajectory followed by the mass center of the body during free fall is a function of the forces applied to it prior to the projection. To achieve maximum translation during the period of free fall, a

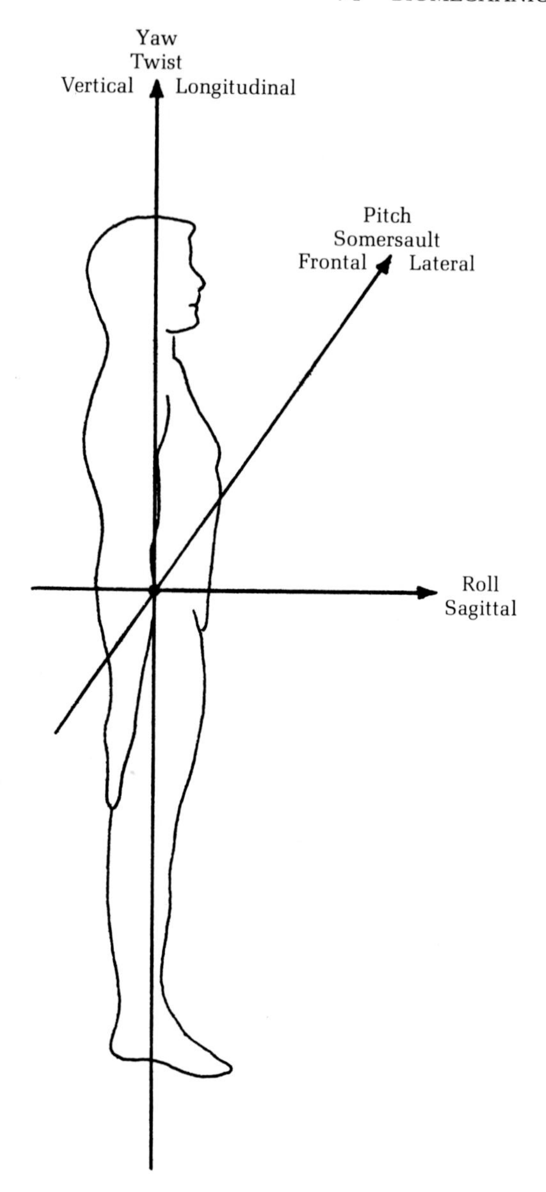

FIGURE 4-20. *Rotational Axes of the Body.*

large force must be directed through the mass center of the body. A force whose line of action passes to one side of the mass center will generate rotation as well as translation. The greater the distance of the force from the mass center, the greater will be its moment or rotary effect.

Assuming that air resistance is negligible, the only force acting upon the body in the unsupported condition is its weight, which is directed vertically downward (in the z direction) from its mass center. Since no forces influence the body horizontally, its acceleration is zero in the x and y (horizontal) directions. Consequently, the horizontal velocity with which the body is projected remains unchanged.

$$\Sigma F_x = ma_x \qquad \Sigma F_y = ma_y$$
$$0 = ma_x \qquad 0 = ma_y$$
$$a_x = 0 \qquad a_y = 0$$
$$v_x = \int a_x\, dt \qquad v_y = \int a_y\, dt$$
$$= a_x t + C \qquad = a_y t + C$$
$$\text{when } t = 0,\ v_x = v_{x_i} \qquad \text{when } t = 0,\ v_y = v_{y_i}$$
$$C = v_{x_i} \qquad C = v_{y_i}$$
$$v_x = v_{x_i} \qquad v_y = v_{y_i}$$

in which x and y indicate the two orthogonal horizontal components; *i* is the initial value; *t* and *dt* indicate time; and *C* is the constant of integration. Continuing the integration:

$$s_x = \int v_x\, dt \qquad s_y = \int v_y\, dt$$
$$= \int v_{x_i}\, dt \qquad = \int v_{y_i}\, dt$$
$$= v_{x_i} t + C \qquad = v_{y_i} t + C$$
$$\text{when } t = 0,\ s_x = s_{x_i} \qquad \text{when } t = 0,\ s_y = s_{y_i}$$

in which s represents displacement and v indicates velocity. As shown by the equations, the horizontal components of velocity remain unchanged as long as the body is in the air.

The body, however, does have a limited time during which it is airborne. This is determined by the magnitude of the vertical force with which it is projected. Once free in the air, the body is pulled back toward the ground by the gravitational attraction of the earth. Such a force results in a constant acceleration of all parts of the body of −32.2 ft/sec/sec.†

† Since the acceleration due to gravity is influenced by the distance to the center of the earth, its precise magnitude may vary slightly from one geographic location to another.

$$\Sigma F_z = ma_z$$

$$-W = \frac{W}{g} a_z$$

$$a_z = -g = -32.2 \text{ ft/sec/sec}$$

$$v_z = \int a_z \, dt$$

$$= a_z t + C$$

$$\text{when } t = 0, v_z = v_{z_i}$$

$$C = v_{z_i}$$

$$v_z = a_z t + v_{z_i}$$

$$s_z = \int v_{z_i} \, dt$$

$$= \int (a_z t + v_{z_i}) \, dt$$

$$s_z = \tfrac{1}{2} a_z t^2 + v_z t + C$$

$$\text{when } t = 0, s_z = s_{z_i}$$

$$C = s_{z_i}$$

$$s_z = \tfrac{1}{2} a_z t^2 + v_{z_i} t + s_{z_i}$$

These equations provide the means for determining the velocity or displacement components of a freely falling body at any point in time provided its velocity and displacement values at the instant of projection are known. Their resultants can also be calculated from the relationships:

$$s = \sqrt{s_x^{\,2} + s_y^{\,2} + s_z^{\,2}} \qquad v = \sqrt{v_x^{\,2} + v_y^{\,2} + v_z^{\,2}}.$$

Although the translation of the mass center initially involves a kinetic analysis, once the equations have been derived the problem reduces to one of kinematics. Therefore, it is not necessary to know the magnitudes of the forces to be able to predict the characteristics of the trajectory. From a practical point of view, this means that the path of the performer's mass center is established at the take-off. Once he has broken contact with his supporting surface, he cannot alter his trajectory even though he may be able to move his body segments with respect to his mass center. Thus, the hitch kick action of the long jumper does not alter the path of his mass center although it does have implications for the control of body rotation. Similarly, the trampolinist who initiates a somersault with excessive horizontal velocity will probably land on the floor in spite of any frantic body gyrations which he may make in the air.

What then is the optimum angle of projection for the athlete or his sports implement? The answer is not so simple as it might appear. The first, and most important consideration, is the objective of the performance. Is it to achieve height, distance, rotation, accuracy or some combination of two or more of these factors? When the goal is maximum vertical displacement of the body or time in the air, as would be the case in a volleyball block or basketball rebound, the greatest ground reaction force possible should be directed vertically upward through the mass center. If horizontal distance is desired such as in the shot put or long jump, the line of action of the force should again pass as close to the mass center of the projectile as possible. In this case, both horizontal and vertical components of force are required since the body must have sufficient time in the air to permit unimpeded horizontal movement.

If the projection and landing levels were the same and the performer were capable of generating identical magnitudes of force throughout his range of motion, then a projection angle of 45 degrees would achieve the greatest horizontal displacement. In sport, however, this is seldom the case. The level of projection is usually higher than that of the landing and this favors a lower projection angle. Further, the human body is not capable of applying equal levels of force at all projection angles. As the angle increases above the horizontal, additional work must be done to overcome the resistance of the weight. Thus, the task becomes more difficult.

All of these factors tend to favor projection angles less than the frequently quoted 45 degrees. Therefore, in estimating the optimum projection angle for a specific performance, consideration must be given to the objective of the activity, muscular force variations throughout the range of motion, and the height of projection with respect to the landing level. Air or wind resistance must also be taken into account if they markedly affect the outcome.

Tables 4-5 and 4-6 illustrate the results of the flight of a shot as a function of the release height and the magnitude and direction of its velocity at release. The horizontal distance achieved in each case was calculated from the equations of motion of a particle in free fall derived earlier in this section. Even though the shot put provides one of the simplest examples of free fall in sport, careful examination of the data will furnish information of practical significance to the teacher and coach.

From the theoretical calculations presented, certain conclusions appear warranted. First, the height of the shot at release is directly related to the horizontal distance achieved in the put. If realistic projection velocities are used, a shot released eight feet above the ground will travel about a foot farther than one projected from a

Table 4-5. Computer Simulation of the Horizontal Distance Covered by a Shot as a Function of Velocity and Height of Release

Velocity of Release				*Distance Thrown as a Function of Height of Release*				
Resultant	*Horizontal (Feet/Second)*	*Vertical*	*Angle (Degrees)*	6	6.5	7 (Feet)	7.5	8
28.00	24.25	14.00	30.00	28.72	29.21	29.70	30.17	30.63
28.00	24.00	14.42	31.00	28.92	29.41	29.88	30.34	30.79
28.00	23.75	14.84	32.00	29.10	29.58	30.04	30.50	30.94
28.00	23.48	15.25	33.00	29.27	29.73	30.19	30.63	31.06
28.00	23.21	15.66	34.00	29.40	29.86	30.31	30.74	31.17
28.00	22.94	16.06	35.00	29.52	29.97	30.40	30.83	31.25
28.00	22.65	16.46	36.00	29.61	30.05	30.48	30.89	31.30
28.00	22.36	16.85	37.00	29.68	30.11	30.53	30.93	31.33
28.00	22.06	17.24	38.00	29.73	30.14	30.55	30.95	31.34
28.00	21.76	17.62	39.00	29.75	30.16	30.55	30.94	31.33
28.00	21.45	18.00	40.00	29.74	30.14	30.53	30.91	31.29
28.00	21.13	18.37	41.00	29.71	30.10	30.48	30.85	31.22
28.00	20.81	18.74	42.00	29.66	30.03	30.41	30.77	31.13
28.00	20.48	19.10	43.00	29.57	29.94	30.30	30.66	31.01
28.00	20.14	19.45	44.00	29.46	29.82	30.18	30.52	30.86
28.00	19.80	19.80	45.00	29.33	29.68	30.02	30.36	30.69
28.00	19.45	20.14	46.00	29.17	29.51	29.84	30.17	30.50
28.00	19.10	20.48	47.00	28.98	29.31	29.64	29.96	30.27
28.00	18.74	20.81	48.00	28.76	29.09	29.40	29.72	30.02
28.00	18.37	21.13	49.00	28.52	28.84	29.14	29.45	29.75
28.00	18.00	21.45	50.00	28.25	28.56	28.86	29.15	29.44
29.00	25.11	14.50	30.00	30.36	30.87	31.36	31.85	32.32
29.00	24.86	14.94	31.00	30.59	31.09	31.57	32.04	32.51
29.00	24.59	15.37	32.00	30.79	31.28	31.76	32.22	32.67
29.00	24.32	15.79	33.00	30.98	31.45	31.92	32.37	32.82
29.00	24.04	16.22	34.00	31.13	31.60	32.06	32.50	32.94
29.00	23.76	16.63	35.00	31.27	31.72	32.17	32.61	33.03
29.00	23.46	17.05	36.00	31.38	31.82	32.26	32.68	33.10
29.00	23.16	17.45	37.00	31.46	31.90	32.32	32.74	33.15
29.00	22.85	17.85	38.00	31.52	31.94	32.36	32.77	33.17
29.00	22.54	18.25	39.00	31.55	31.96	32.37	32.77	33.16
29.00	22.22	18.64	40.00	31.55	31.96	32.35	32.74	33.12
29.00	21.89	19.03	41.00	31.53	31.92	32.31	32.69	33.06
29.00	21.55	19.40	42.00	31.47	31.86	32.24	32.61	32.97
29.00	21.21	19.78	43.00	31.39	31.77	32.14	32.50	32.86
29.00	20.86	20.15	44.00	31.29	31.65	32.01	32.37	32.71
29.00	20.51	20.51	45.00	31.15	31.51	31.86	32.20	32.54
29.00	20.15	20.86	46.00	30.98	31.33	31.67	32.01	32.34
29.00	19.78	21.21	47.00	30.79	31.13	31.46	31.79	32.11
29.00	19.40	21.55	48.00	30.57	30.90	31.22	31.54	31.85
29.00	19.03	21.89	49.00	30.31	30.63	30.95	31.26	31.56
29.00	18.64	22.22	50.00	30.03	30.34	30.65	30.95	31.25

Table 4-5 *Continued*

Velocity of Release				*Distance Thrown as a Function of Height of Release*				
Resultant	*Horizontal (Feet/Second)*	*Vertical*	*Angle (Degrees)*	6	6.5	7 (Feet)	7.5	8
36.00	31.18	18.00	30.00	43.23	43.81	44.38	44.93	45.48
36.00	30.86	18.54	31.00	43.66	44.23	44.78	45.32	45.86
36.00	30.53	19.08	32.00	44.06	44.61	45.15	45.68	46.20
36.00	30.19	19.61	33.00	44.42	44.96	45.48	46.00	46.51
36.00	29.85	20.13	34.00	44.74	45.26	45.78	46.28	46.78
36.00	29.49	20.65	35.00	45.02	45.53	46.03	46.53	47.01
36.00	29.12	21.16	36.00	45.26	45.76	46.25	46.73	47.21
36.00	28.75	21.67	37.00	45.46	45.95	46.43	46.90	47.36
36.00	28.37	22.16	38.00	45.63	46.10	46.57	47.02	47.48
36.00	27.98	22.66	39.00	45.75	46.21	46.66	47.11	47.55
36.00	27.58	23.14	40.00	45.82	46.27	46.72	47.15	47.58
36.00	27.17	23.62	41.00	45.86	46.29	46.73	47.15	47.57
36.00	26.75	24.09	42.00	45.85	46.27	46.69	47.11	47.51
36.00	26.33	24.55	43.00	45.79	46.21	46.62	47.02	47.41
36.00	25.90	25.01	44.00	45.69	46.10	46.49	46.89	47.27
36.00	25.46	25.46	45.00	45.55	45.94	46.33	46.71	47.09
36.00	25.01	25.90	46.00	45.36	45.74	46.12	46.49	46.86
36.00	24.55	26.33	47.00	45.13	45.50	45.86	46.23	46.58
36.00	24.09	26.75	48.00	44.85	45.21	45.56	45.92	46.26
36.00	23.62	27.17	49.00	44.53	44.88	45.22	45.56	45.90
36.00	23.14	27.58	50.00	44.16	44.50	44.83	45.16	45.49
37.00	32.04	18.50	30.00	45.27	45.86	46.43	47.00	47.55
37.00	31.72	19.06	31.00	45.74	46.31	46.87	47.42	47.96
37.00	31.38	19.61	32.00	46.16	46.72	47.27	47.81	48.33
37.00	31.03	20.15	33.00	46.55	47.09	47.63	48.15	48.67
37.00	30.67	20.69	34.00	46.90	47.43	47.95	48.46	48.97
37.00	30.31	21.22	35.00	47.20	47.72	48.23	48.73	49.22
37.00	29.93	21.75	36.00	47.47	47.97	48.47	48.96	49.44
37.00	29.55	22.27	37.00	47.69	48.18	48.67	49.15	49.61
37.00	29.16	22.78	38.00	47.87	48.35	48.82	49.29	49.74
37.00	28.75	23.28	39.00	48.01	48.47	48.93	49.39	49.83
37.00	28.34	23.78	40.00	48.09	48.55	49.00	49.44	49.87
37.00	27.92	24.27	41.00	48.14	48.58	49.02	49.45	49.87
37.00	27.50	24.76	42.00	48.14	48.57	48.99	49.41	49.82
37.00	27.06	25.23	43.00	48.09	48.51	48.92	49.33	49.73
37.00	26.62	25.70	44.00	47.99	48.40	48.80	49.20	49.59
37.00	26.16	26.16	45.00	47.85	48.24	48.63	49.02	49.40
37.00	25.70	26.62	46.00	47.66	48.04	48.42	48.80	49.17
37.00	25.23	27.06	47.00	47.42	47.79	48.16	48.52	48.88
37.00	24.76	27.50	48.00	47.13	47.49	47.85	48.21	48.56
37.00	24.27	27.92	49.00	46.79	47.15	47.50	47.84	48.18
37.00	23.78	28.34	50.00	46.41	46.75	47.09	47.43	47.76

Table 4-6. Shot Put Distance as a Function of the Conditions at Release

Resultant Release Velocity (ft/sec)	*Height of Release*					
	Six ft		*Seven ft*		*Eight ft*	
	Optimum Angle (°)	*Distance (ft)*	*Optimum Angle (°)*	*Distance (ft)*	*Optimum Angle (°)*	*Distance (ft)*
26	38–39	26.32	37–38	27.10	37	27.87
28	39	29.75	38–39	30.55	38	31.34
30	40	33.42	39	34.24	39	35.05
32	40–41	37.32	40	38.16	39	38.99
34	41	41.47	40–41	42.32	39–40	43.16
36	41	45.86	41	46.73	40	47.58
38	42	50.49	41	51.37	40–41	52.23
40	42	55.36	41–42	56.25	41	57.13
42	42	60.48	42	61.38	41	62.27
44	42	65.85	42	66.76	42	67.65
46	43	71.46	42	72.37	42	73.28

height of seven feet. Similarly, it will have roughly an advantage of two feet over a similar put with a release height of six feet. This evidence confirms the natural superiority which taller athletes enjoy over their shorter opponents in the shot put. Secondly, the optimum angle of projection is approximately 41 to 42 degrees above the horizontal. Poorer performers, characterized by their smaller release velocities, have a slightly lower optimum release angle than do top caliber putters. In addition, performers with a higher point of release also achieve a greater horizontal distance from a lower projection angle than athletes using a lower point of release. Finally, the magnitude of the release velocity is extremely important to the success of the performance. In fact, even a cursory evaluation of the data indicates that it is the most important factor. Thus in coaching, increasing the release velocity should be stressed even if it is achieved at the expense of a high point of release and optimum projection angle (Miller, 1973, In Press).

Moving from the translation phase of free fall in sport to the rotation is the next step in a kinetic analysis. Once the performer is free in the air not only is the trajectory fixed but also the angular momentum with respect to the mass center. The latter is explained by the principle which states that the angular momentum with respect to the mass center remains constant unless there is a corresponding angular impulse about the mass center.

Angular impulse of a force **F** about an axis through point o is defined as:

$$(\mathbf{AI})_o = \int \mathbf{r} \times \mathbf{F}\, dt$$

which is simply the moment of the linear impulse

$$\mathbf{LI} = \int \mathbf{F}\, dt.$$

(The designation **r** refers to a position vector joining the axis to the line of action of **F**.) In the special case where point o is the mass center (g) of a rigid body, the angular impulse is equal to the change in the angular momentum (**H**). Thus,

$$(\mathbf{AI})_g = \Delta \mathbf{H}_g.$$

In the free fall instance where air resistance is disregarded, the only force acting upon the system is its weight. Since the latter is applied through the mass center, it can have no impulse with respect to g. Therefore, there cannot be any change in the total angular momentum while this condition exists. This means that the angular momentum which the athlete established at take-off remains constant during the flight portion of his performance. For a multi-link interconnected system such as the human body, there may be changes in the angular momentum of the individual segments but the resultant angular momentum of the whole system will not change.

Angular momentum or the moment of momentum as it is sometimes called refers to the "quantity" of angular motion. The angular momentum of a mass particle m_i about an axis through point o is expressed as:

$$\mathbf{H}_o = \mathbf{r}_i \times m_i \mathbf{v}_i$$

which is clearly the moment of the linear momentum, $\mathbf{G} = m_i \mathbf{v}_i$. The angular momentum of a rigid body is the sum of the moments of the linear momenta of all its composite particles. When the rigid body undergoes planar motion, the magnitude of its angular momentum is (see Meriam, 1966):

$$H_o = I_g \omega + m v_g\, d$$

in which H_o is the angular momentum with respect to point o;
I_g is the moment of inertia of the body with respect to the mass center;
v_g is the velocity of the mass center;
ω is the angular velocity of the body;
d is the perpendicular distance from point o to vector $\mathbf{v_g}$; and
m is the mass.

The angular momentum vector is normal to the plane of the motion in the direction indicated by the right hand rule.† When o is a fixed point, the angular momentum relationship reduces to the following:

$$\begin{aligned} H_o &= I_g\omega + mv_g\, d \\ &= (I_o - m\, d^2)\omega + m(\omega\, d)\, d \\ &= I_o\omega - m\omega\, d^2 + m\omega\, d^2 \\ &= I_o\omega. \end{aligned}$$

Similarly, if o is the mass center:

$$\begin{aligned} H_g &= I_g\omega + mv_g\, d \\ &= I_g\omega + mv_g 0 \\ &= I_g\omega. \end{aligned}$$

At a rather superficial level of analysis, the body can be considered in a series of quasi-rigid positions during the unsupported phase of the skill. Since the angular momentum remains constant with reference to the mass center,

$$H_g = I_g\omega = mk_g^{\,2}\omega = \text{a constant}$$

in which k represents the radius of gyration. This can be further simplified to

$$k^2\omega = \text{a constant}$$

because the mass of the body does not change during the course of the motion. An increase in the radius of gyration causes a corresponding decrease in the angular velocity of the body and vice versa. It is commonly observed that a gymnast spins faster in a tuck position than in a layout. Similarly, to execute a multiple rotation jump such as a double axel, the figure skater must be in an upright position with the arms and free leg as close to the body as possible to reduce the radius of gyration with respect to the longitudinal axis. Thus, an athlete can control his rate of rotation (angular velocity) by adjusting the distribution of his body mass about the rotational axis. The total angular momentum about an axis through the mass center, however, remains the same.

A more rigorous kinetic analysis based upon principles of angular momentum treats the human body as a linked system. Since the complexity of the relationship is a positive function of the number of segments, it is advisable to represent the body as simply as possible. For example, if both legs remain straight and together during the performance, they can be treated as a single segment rather than six individual links. In addition, the trunk and head may sometimes be considered as one rigid body. If the performer, designated system

†Refer to Appendix B.

S, is represented by four rigid segments as follows: the head-trunk ($B1$), the legs ($B2$), left arm ($B3$) and right arm ($B4$), his angular momentum (**H**) with respect to its mass center (CS) is (see Smith and Kane, 1967; Miller, 1970):

$$\mathbf{H}^{S/CS} = \mathbf{H}^{B1/C1} + \mathbf{H}^{B2/C2} + \mathbf{H}^{B3/C3} + \mathbf{H}^{C4/C4} + \mathbf{H}^{C1/CS} + \mathbf{H}^{C2/CS} + \mathbf{H}^{C3/CS} + \mathbf{H}^{C4/CS}.$$

The first four terms on the right hand side of the equation represent the angular momentum ($I\omega$) of each rigid body with respect to its own mass center. In the spatial case, these terms are expressed as three orthogonal angular momentum components about the principal axes of the segment. The final four terms in the total angular momentum relationship indicate the moment of the linear momentum of each segmental mass center with reference to the mass center of the whole system.

Thus, the angular momentum of each segment as well as the moment of the linear momentum of each segmental mass center contributes to the total angular momentum of the body. Since the latter remains constant during the unsupported phase of the skill, alteration in the motion of one segment is reflected in the motion of the others (Figure 4-21). If a long jumper leaves the take-off board with excessive angular momentum which tends to somersault him forward, he can control this undesired rotation by circling the arms forward and by performing a hitch kick. If the angular momentum contributed by the limbs equals the total angular momentum initiated at the take-off, the head and trunk will have a zero angular velocity and will remain in their take-off position. Should the limbs generate more angular momentum than the total established initially, then the trunk would have to assume the appropriate amount of negative angular velocity or backward rotation in order to maintain the constant angular momentum of the whole body. Similarly, the diver, performing a reverse two and one-half somersault tuck, who prepares for the entry by swinging his arms straight back over his head in the direction of the spin actually slows the rotation of his trunk and legs. This action not only increases the moment of inertia with respect to the spinning axis but "uses up" some of the total angular momentum of the body so there is less to be accounted for by the remaining segments. Accordingly, the trampolinist executing a back somersault layout leaves the bed with his arms extended above his head and subsequently pulls them down by his sides in a direction completely opposite to the rotation of the body. This arm action results in an increased angular velocity of the trunk and legs because it reduces the resistance to angular rotation. It also sets up a negative angular momentum which must be compensated for by

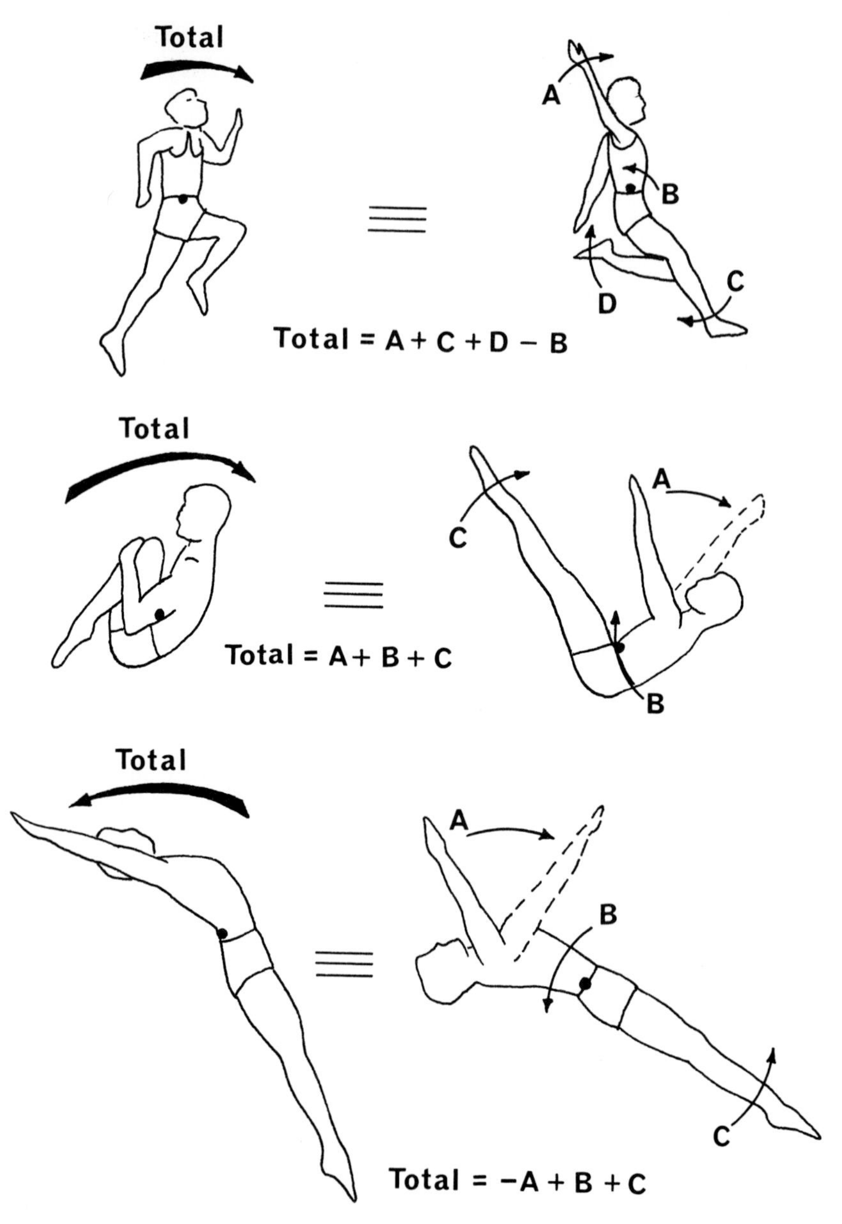

FIGURE 4-21. *Individual Segment Contributions to Total Angular Momentum.*

an increased positive angular momentum of the remainder of the body. Thus, the somersault is easier to complete.

The influence of the motion and position of the individual segments upon the remainder of the body in a condition of free fall is, therefore, dependent upon the terms in the angular momentum relationship for a linked system. The relative magnitudes of segmental masses and moments of inertia are indicated in Table 4-7. Also important are the displacements and velocities of the limbs. Only a quantitative kinetic analysis, however, can provide specific data on the effect of a single movement upon the total performance of a skill.

Table 4-7. Segmental Parameters of an Adult Male Subject Generated by Hanavan's Computer Model†

Segment	*Weight* (*lb*)	*Mass* (*Slugs*)	*Length* (*in*)	*Principal Moments of Inertia (Slug-in-in)*		
				Ix	*Iy*	*Iz*
Head	12.10	0.38	12.82	4.03	4.03	1.88
Upper Trunk	16.98	0.53	6.00	6.40	3.82	7.05
Lower Trunk	55.85	1.73	16.76	58.03	48.29	25.12
Hand	1.13	0.04	3.54	0.04	0.04	0.04
Upper Arm	4.73	0.15	12.10	1.90	1.90	0.24
Lower Arm	2.85	0.09	10.01	0.75	0.75	0.09
Thigh	15.56	0.48	11.66	6.24	6.24	2.14
Lower Leg	7.56	0.23	15.95	4.94	4.94	0.41
Foot	2.31	0.07	10.20	0.62	0.62	0.06

†Refer to Chapter 5, pages 112–113 for further information on this model.

SUMMARY

Since purposeful motion is the essence of sport, the principles of dynamics must be applied in evaluating this motion quantitatively. Initially, a temporal analysis can be utilized to identify the rhythm of the various movement components. The second level of investigation, referred to as kinematics, includes a careful examination of the time-geometry of motion. Linear and angular displacement, velocity and acceleration of the body segments are studied in relation to the objectives of the performance. Finally, the forces responsible for the movement can be investigated. This is known as a kinetic analysis and represents the most detailed and complex aspect of dynamics.

SELECTED REFERENCES

Atwater, A. E.: Movement Characteristics of the Overarm Throw: A Kinematic Analysis of Men and Women Performers. Unpublished Doctoral Dissertation, University of Wisconsin, 1970.

Batterman, C.: *The Techniques of Springboard Diving*. Cambridge, Mass.: MIT Press, 1968.

Carlsöö, S.: A Kinetic Analysis of the Golf Swing. J. Sports Med., *7*, 76–82, 1967.

Cavagna, G. A., Komarek, L., and Mazzoleni, S.: The Mechanics of Sprint Running. J. Physiol. (Lond.), *217*, 709–721, 1971.

Cavagna, G. A., Margaria, R., and Arcelli, E.: A High-Speed Motion Picture Analysis of the Work Performed in Sprint Running. Res. Film, *5*, 309–319, 1965.

Cavagna, G. A., Saibene, F. P., and Margaria, R.: Mechanical Work in Running. J. Appl. Physiol., *19*, 249–256, 1964.

Chapman, S.: Catching a Baseball. Amer. J. Phys., *36*, 868–870, 1968.

Dillman, C. J.: A Kinetic Analysis of the Recovery Leg During Sprint Running. In J. M. Cooper (Ed.), *Selected Topics on Biomechanics*. Chicago: Athletic Institute, 1971.

Dillman, C. J.: Muscular Torque Patterns of the Leg During the Recovery Phase of Sprint Running. Unpublished Doctoral Dissertation, Pennsylvania State University, 1970.

Dillman, C. J., and Nelson, R. C.: The Mechanical Energy Transformations of Pole Vaulting with a Fiberglass Pole. J. Biomech., *1*, 175–183, 1968.

Drillis, R., and Contini, R.: Body Segment Parameters. School of Engineering and Science, New York University, 1966 (PB 174 945; Tech. Rpt. No. 1166.03).

Eaves, G.: *Diving—The Mechanics of Springboard and Firmboard Techniques*. London: Kaye & Ward, 1969.

Elftman, H.: The Work Done by Muscles in Running. Amer. J. Physiol., *129*, 672–684, 1940.

Elftman, H.: The Function of Muscles in Locomotion. Amer. J. Physiol., *125*, 357–366, 1939a.

Elftman, H.: Forces and Energy Changes in the Leg During Walking. Amer. J. Physiol., *125*, 339–356, 1939b.

Falize, J. L., Lucassen, J-P., and Hunebelle, G.: Analyse de l'Impulsion dans le Saut en Hauteur sans Elan. Kinanthropologie, *1*, 25–43, 1969.

Fenn, W. O.: Work Against Gravity and Work due to Velocity Changes in Running. Amer. J. Physiol., *93*, 433–462, 1930a.

Fenn, W. O.: Frictional and Kinetic Factors in the Work of Sprint Running. Amer. J. Physiol., *92*, 583–611, 1930b.

Gerrish, P. H.: *A Dynamic Analysis of the Standing Vertical Jump*. New York: Columbia University, 1934.

Hanavan, E. P.: A Mathematical Model of the Human Body. Wright-Patterson AFB, Ohio, 1964 (AMRL-TR-64-102).

Hay, J. G.: Mechanical Energy Relationships in Vaulting with a Fibreglass Pole. Ergonomics, *14*, 437–448, 1971.

Hay, J. G.: An Investigation of Mechanical Efficiency in Two Styles of High Jumping. Unpublished Doctoral Dissertation, University of Iowa, 1967.

Hunebelle, G., and Damoiseau, J.: Relations Between Performance in High Jump and Graph of Impulsion. Paper presented at the Third International Seminar on Biomechanics, Rome, 1971.

Jorgensen, T.: On the Dynamics of the Swing of a Golf Club. Amer. J. Phys., *38*, 644–651, 1970.

Kirkpatrick, P.: Batting the Ball. Amer. J. Phys., *31*, 606–613, 1963.

Kuhlow, A.: *Analyse moderner Hochsprungtechniken—Ein Beitrag zur Speziellen Bewegungslehre der Leibesübungen*. Berlin: Bartels & Wernitz, 1971.

Magel, J. R.: Propelling Force Measured During Tethered Swimming in the Four Competitive Swimming Styles. Res. Q. Amer. Assoc. Health Phys., Ed., *41*, 68–74, 1970.

Meriam, J. L.: *Dynamics*. New York: Wiley, 1966.

Miller, D. I.: Computer Simulation of Human Motion. In H. T. A. Whiting (Ed.), *Techniques for the Analysis of Human Movement*. London: Henry Kimpton, 1973, In Press.

Miller, D. I.: A Computer Simulation Model of the Airborne Phase of Diving. In J. M. Cooper (Ed.), *Selected Topics on Biomechanics*. Chicago: Athletic Institute, 1971.

Miller, D. I.: A Computer Simulation Model of the Airborne Phase of Diving. Unpublished Doctoral Dissertation, Pennsylvania State University, 1970.

Murray, M. P., Seireg, A., and Scholz, R. C.: Center of Gravity, Center of Pressure, and Supportive Forces during Human Activities. J. Appl. Physiol., *23*, 831–838, 1967.

Offenbacher, E. L.: Physics and the Vertical Jump. Amer. J. Phys., *38*, 829–836, 1970.

Payne, A. H., Slater, W. J., and Telford, T.: The Use of a Force Platform in the Study of Athletic Activities. A Preliminary Investigation. Ergonomics, *11*, 123–143, 1968.

Pearson, J. R., McGinley, D. R., and Butzel, L. M.: A Dynamic Analysis of the Upper Extremity: Planar Motions. Hum. Factors, *5*, 59–70, 1963.

Plagenhoef, S.: *Patterns of Human Motion—A Cinematographic Analysis*. Englewood Cliffs, N.J.: Prentice-Hall, 1971.

Plagenhoef, S. C.: Methods for Obtaining Kinetic Data to Analyze Human Motions. Res. Q. Amer. Assoc. Health Phys. Ed., *37*, 103–112, 1966.

Plagenhoef, S. C.: An Analysis of the Kinematics and Kinetics of Selected Symmetrical Body Actions. Unpublished Doctoral Dissertation, University of Michigan, 1962.

Ramey, M. R.: Effective Use of Force Plates for Long Jump Studies. Res. Q. Amer. Assoc. Health Phys. Ed., *43*, 247–252, 1972.

Ramey, M. R.: Force Relationships of the Running Long Jump. Med. Sci. Sports, *2*, 146–151, 1970.

Shames, I. H.: *Engineering Mechanics—Statics and Dynamics*. 2nd Ed., Englewood Cliffs, N.J.: Prentice-Hall, 1967.

Sinning, W. E., and Forsyth, H. L.: Lower-limb Actions While Running at Different Velocities. Med. Sci. Sports, *2*, 28–34, 1970.

Smith, P. G., and Kane, T. R.: The Reorientation of a Human Being in Free Fall. Division of Engineering Mechanics, Stanford University, 1967 (Tech. Rpt. No. 171).

Williams, D.: The Dynamics of the Golf Swing. Quart. J. Mech. Appl. Math., *20*, 247–264, 1967.

CHAPTER 5

Body Segment Parameters

THE HUMAN ORGANISM seldom acts as a single rigid body whose motion is represented by the displacement of its mass center. It is more accurately described as an n-link mechanical system with multiple degrees of freedom at the junctures of the segments. While, admittedly, this description incorporates many simplifying assumptions, it is extremely useful from an analytical point of view. However, if quantitative evaluations of the dynamics of human motion in sport are to be undertaken, estimates of the lengths, masses, centers of mass and moments of inertia of these body segments must be available.

In keeping with the mechanical analogy, segments are commonly referred to as links. The length of an anatomical link has been defined by Dempster (1955a) as a straight core or axial line connecting two adjacent joint centers or, in the case of a terminal segment, the straight line between the joint center and the mass center of the segment. These terms of reference, however, do not facilitate the measurement of link lengths because the functional joint centers do not correspond with surface landmarks. Clauser and his associates (1969) have therefore suggested that segment lengths be determined by utilizing bony landmarks which can be readily identified.

The mass of an individual segment, which represents its inertia in quantitative terms, is a function of its volume and density. The latter is influenced by the relative amounts of bone, muscle and adipose tissue composing the segment. Although the mass of a human limb is not strictly uniform throughout its length, such an assumption is often entertained to facilitate calculations.

Each segment is composed of an almost infinite number of mass particles. For convenience of mechanical analysis, the resultant of all of these mass particles is considered to act at one point—the center of mass or center of gravity of the segment. Implicit in this concept is the principle that the weight, acting at the mass center, is equal to the sum of the weights of all the mass particles and that the individual particle weights balance themselves about the mass center. Thus, if a segment were suspended at this point, it would remain in equilibrium and there would be no rotation. If a body were uniform in density throughout its length, as is characteristic of many engineering systems, then the center of mass and the center of volume would be coincident. This, in fact, is an assumption which is sometimes held in the case of the segments of the human body.†

The most difficult of the body segment parameters to obtain are the moments of inertia, the quantitative measures of a body's resistance to changes in angular acceleration. A knowledge and an understanding of these parameters, however, are essential in most calculations involving angular motion, and this includes virtually all human movement in sport. The magnitude of a moment of inertia is a function of the mass of the segment and the distribution of that mass from a particular axis of rotation. The estimation of these human segmental parameters falls into three principal classifications: cadaver studies, techniques used on living subjects and mathematical models.

CADAVER STUDIES

The difficulty of acquiring large representative samples of various populations for dissection is evidenced by the fact that over the past century the body segment parameters of fewer than 50 cadavers have been studied. In addition, this type of investigation has been largely restricted to adult males of the Caucasian race. Much of the initial

†Clauser, McConville and Young (1969), however, have shown that the center of mass of the limbs is not coincident with the center of volume but somewhat distal to it.

research was conducted in Germany in the latter half of the nineteenth century. Subsequently, the work was pursued in Japan and in the United States.

One of the earliest detailed studies of human body segment parameters was made by Harless (1860). He dismembered two adult male cadavers and then weighed and measured the body parts. Segmental mass centers were located by using a balancing method developed earlier by the Weber brothers. Harless' work preceded by almost a century modern attempts to describe the human body mathematically. He suggested that the upper part of the torso could be represented by a truncated cone and the lower part by a space surrounded by an elliptical surface. Employing anthropometric measurements from the cadaver, standard mathematical volume formulae and a specific gravity factor of 1.066, he estimated the masses of the two parts of the trunk. The total mass compared favorably with that determined experimentally for the whole torso. In conjunction with his interest in predicting segmental mass distribution in living subjects, Harless also investigated the specific gravities of the body segments of two male and two female cadavers.

With a sample of four adult male cadavers, all suicide victims, Braune and Fischer (1889) conducted one of the classical studies in the field of body segment parameters. Unfortunately, one of the specimens could not be dissected, thereby reducing the sample size to three. Both total body and segmental mass center locations were determined. They concluded that the mass centers of extremities could be found approximately four-ninths of the way down the limb from the proximal joint. Later, Fischer (1906) reported the segmental weights, masses and centers of mass of a single cadaver. He also computed the segmental moments of inertia with respect to an axis through the mass center and perpendicular to the long axis.

The most widely cited and comprehensive investigation of body segment parameters was conducted by Dempster (1955a). In carefully dissecting eight male cadavers, he not only furnished important data but also perfected and standardized research techniques in this field. Segmental centers of gravity were determined with a balance plate; moments of inertia by setting the segment up as a free swinging pendulum system and calculating the period of oscillation; and volumes using the method of immersion. Dempster's data have been used extensively in the development of mathematical models of the human body and in the kinetic analysis of man's motion.

Mori and Yamamoto (1959) and Fujikawa (1963) performed dissections on Japanese cadavers to determine segmental mass proportions and center of gravity locations, respectively. In both investigations, a sample size of six including both male and female specimens was studied. Differing racial influences upon body structure, however, do not permit these data to be statistically combined with those of Caucasian subjects.

Clauser and his associates (1969), utilizing techniques similar to those of Braune and Fischer (1889) and Dempster (1955a), studied the segmental properties of 14 male cadavers. Since the first specimen was used to perfect dissection techniques and standardize procedures, only data on the other 13 were reported. Anthropometric measurements were made and somatotype photographs taken. After the total body volume and center of mass had been ascertained, the segments were severed and their weights, volumes and mass centers established. Stepwise regression equations were then developed for predicting these three segmental parameters from anthropometric measurements.

Although the trunk is commonly portrayed as a rigid body, it is generally appreciated that such a representation is not very accurate. Concern with the segmental properties of this portion of the body has been expressed by several investigators. Parks (1959) carried out a detailed analysis of the trunk of a single male cadaver. He first divided it into thoracic, abdominal and pelvic segments and subsequently located the mass centers of eight thoracic and five abdominal cross sections. More recently, Liu and his associates (1971) sectioned a cadaver trunk into 25 horizontal slices. The mass of each was calculated by weighing it and then dividing by the acceleration due to gravity. Balancing each section in three different positions on a knife edge gave information on the location of its mass center. A torsion pendulum was utilized to obtain the principal moments of inertia. Problems were encountered, however, in making precise cuts and some error was introduced by tissue shift prior to sectioning. The limited nature of these two attempts to provide more information on the segmental properties of the trunk is evident. The difficulty in obtaining such data is also appreciated.

An inspection of Table 5-1 will support the fact that the cadavers studied have not been representative of the average adult population with respect to age, height or weight. Segmental data on Caucasian women are conspicuous by their absence. With the exception of Meeh's work (1895), there is a dearth of information on the body segment characteristics of children. While the majority of research has been devoted to determining segmental masses as a proportion of the total body mass and the location of the mass centers, little has been reported on moments of inertia. A great deal more work is required in this area. The value of dissection studies lies in their potential for providing a basis for prediction of segmental parameters in living subjects. However, because of the nature of the samples

Table 5-1. Summary of Cadaver Studies

Investigators	*Report Publ.*	*Ident.*	*Sex*	*Age (yr)*	*Height (cm)*	*Weight (kg)*	*Build*	*Cause of Death*	*Parameters Studied*
Harless (Germany) N = 2	1860	Graf	Male	—	172.685	63.97		Execution	Segmental length, mass, centers of mass. Segmental specific gravities determined on Kefer and three other cadavers.
		Kefer	Male	29	167.85	49.895		Execution	
Braune and Fischer (Germany) N = 3	1889	1	Male	18	169		well built	Suicide	Total body and segmental centers of mass. Cadaver 1 could not be dissected.
		2	Male	45	170	75.1	muscular	Suicide	
		3	Male	50	166	60.75	muscular	Suicide	
		4	Male	—	168.8	56.09	muscular	Suicide	
Meeh (Germany) N = 4	1895	1	Male	Newborn					Segmental weights and volumes.
		2	Female	8 mo. Stillborn					
		3	Female	Newborn					
		4	Male	1 yr 10 mo.					
Fischer (Germany) N = 1	1906		Male			44.06			Segmental mass, center of gravity, moments of inertia.
Dempster (U.S.A.) N = 8	1955	14815	Male	67	168.9	51.36	somatotype 4–5–2½	Unknown	Anthropometric measurements, segmental masses, centers of mass, density, moments of inertia.
		15059	Male	52	159.8	58.41	3–5–3	Cerebral Hemorrhage	
		15062	Male	75	169.6	58.41	4–2–4	Atherosclerosis	
		15095	Male	83	155.3	49.66	4–3–4	Unknown	
		15097	Male	73	176.64	72.50	4–5–3	Esophageal Carcinoma	
		15168	Male	61	186.6	71.36	3–3–4	Coronary Thrombosis	

		15250	Male	—	180.3	60.45	3–3–4	Acute Coronary Occlusion	
		15251	Male	—	158.5	55.91	4–4–2	Chronic Myocarditis	
Mori and Yamamoto (Japan) N = 6	1959	I	Male	44				Tuberculosis	Segmental masses.
		II	Male	67				Brain Hemorrhage	
		III	Male	85				Old Age	
		IV	Female	37				Cirrhosis of Liver	
		V	Female	57				Stomach Ulcer	
		VI	Female	83				Old Age	
Fujikawa (Japan) N = 6	1963	I	Male	58					Center of gravity of the body segments. Data presented on only five cadavers.
		II	Female	70					
		III	Male	28					
		IV	Female	83					
		V	Male	77					
Clauser and associates (U.S.A.) N = 14	1969		Male	*Range* 28–74 mean 49.31 S.D. 13.69	*Range* 162.5–184.9 172.72 5.94	*Range* 54–87.9 66.52 8.7			Anthropometric measurements, total body and segmental mass, center of mass, volume. Complete data on 13 cadavers. One used to work out techniques. Only group statistics given.
Liu and associates (U.S.A.) N = 1	1971		Male	83	170.18	69.0	medium	Heart Failure	Mass, center of mass, and mass moment of inertia about each of the three principal axes of the sectioned cadaver trunk. Only the trunk was studied.

included, the scope of inference is necessarily limited. Applications have been extended to populations of athletes, military personnel and, in some cases, even women and children. The confidence associated with such predictions cannot be considered high. There is, in addition, the basic question of whether cadaver data can logically be applied to living subjects. While this assumption is often held, it is extremely difficult to validate with humans.

TECHNIQUES WITH LIVING SUBJECTS

The problem of obtaining adequate estimates of the segmental properties of living subjects has confronted biomechanics researchers for a number of years. Direct measurement of the masses, mass centers and moments of inertia of the body segments is quite difficult, particularly in the latter case. Several different methods have been proposed, each with its inherent limitations. Some, while appearing to be theoretically sound, are restricted in application to only a few body segments; others require an excessive amount of time for data collection, necessitate complicated data reduction, lack requisite accuracy, need expensive or intricate custom-designed equipment, or for other reasons are simply not administratively feasible. As yet, no single technique has gained universal acceptance. Thus, the objective of the measurement, the use to which the data will be put and the degree of accuracy desired will substantially influence the particular method or methods chosen. Estimates of body segment parameters have been obtained by such widely divergent means as computation, immersion, reaction change, casting, photogrammetry and, most recently, radiation techniques.

Computation Procedures. The simplest means of approximating segmental masses is to assume fixed relationships exist between the mass of the segments and that of the total body. These relationships can be determined from existing cadaver data. In practice, the subject's body weight is determined. The appropriate segmental weight proportions are then calculated and subsequently converted to units of mass. The earliest attempts to use such computational methods were made by Harless (1860) and Braune and Fischer (1889). Later, Dempster's data (1955a) were employed directly for this purpose. In 1957, Barter statistically combined the data of Braune and Fischer, Fischer (1906) and Dempster to increase the sample size to 12. The regression equations which he developed to predict segment weight as a function of total body weight are presented in Table 5-2. Because of the differences in the dissection techniques employed for the head and neck in the three studies, these results could not be combined.

Table 5-2. Regression Equations for Calculating Segmental Weight in Pounds†

Body Segment		*Regression Equation*	*Standard Error of Estimate*
Head, Neck and Trunk	=	.47 × Total Body Wt. + 12.0	(± 6.4)
Total Upper Extremities	=	.13 × Total Body Wt. − 3.0	(± 2.1)
Both Upper Arms	=	.08 × Total Body Wt. − 2.9	(± 1.0)
Forearms plus Hands	=	.06 × Total Body Wt. − 1.4	(± 1.2)
Both Forearms*	=	.04 × Total Body Wt. − .5	(± 1.0)
Both Hands*	=	.01 × Total Body Wt. + .7	(± 0.4)
Total Lower Extremities	=	.31 × Total Body Wt. + 2.7	(± 4.9)
Both Upper Legs	=	.18 × Total Body Wt. + 3.2	(± 3.6)
Both Lower Legs plus Feet	=	.13 × Total Body Wt. − 0.5	(± 2.0)
Both Lower Legs	=	.11 × Total Body Wt. − 1.9	(± 1.6)
Both Feet	=	.02 × Total Body Wt. + 1.5	(± 0.6)

*N = 11, all others N = 12.

†(Barter, J. T.: Estimation of the Mass of Body Segments. Wright-Patterson Air Force Base, Ohio, p. 6, 1957. WADC TR 57-260).

Barter, therefore, recommended that Dempster's data on the head and neck be used.

The stepwise regression equations of Clauser and his associates (1969) for the prediction of segmental weight from anthropometric measurements including body weight are shown in Table 5-3. Coefficients indicating the correlation of the computed segment weight with the actual one are given along with the standard error of the estimate. It can be seen that total body weight proved to be the single most important variable in predicting segment weight. The inclusion of two or three anthropometric variables in the equations, however, decreased the magnitude of the error. The summary of the segmental weights as simple proportions of total body weight derived from existing studies using adult male Caucasian subjects is presented in Table 5-4.

Fixed relationships can also be assumed between the length of body segments and the location of their mass centers. Table 5-5 presents the findings of Harless (1860), Braune and Fischer (1889), Dempster (1955a) and Clauser and his co-workers (1969). As a general approximation, Braune and Fischer suggested that the mass center of the arm, forearm, thigh, leg and foot could be located a distance four-ninths (44.44%) of the way down the segment from the proximal joint. Dempster indicated the figure to be 43%. These three investigators agreed that the centers of mass of the limbs were located on the long axis between the joint centers. Clauser and his associates

Table 5-3. Regression Equations for Predicting Segmental Weight in Kilograms

				Constant	R	SE Est
Head	*Head Circ*	*Weight (kg)*				
	0.148			−3.716	.814	0.20
	0.104	+0.015		−2.189	.875	0.17
Trunk	*Weight (kg)*	*Trunk Length*	*Chest Circ*			
	0.551			−2.837	.966	1.33
	0.494	+0.347		−19.186	.979	1.11
	0.349	+0.423	+0.229	−35.460	.986	0.92
Upper Arm	*Weight (kg)*	*Arm Circ(ax)*	*Acrom-Rad Lth*			
	0.030			−0.238	.879	0.14
	0.019	+0.060		−1.280	.931	0.12
	0.007	+0.092	+0.050	−3.101	.961	0.09
Forearm	*Wrist Circ*	*Forearm Circ*				
	0.119			−0.913	.827	0.09
	0.081	+0.052		−1.650	.920	0.06
Hand	*Wrist Circ*	*Wrist Br/Bone*	*Hand Brdth*			
	0.051			−0.418	.863	0.03
	0.038	+0.080		−0.660	.917	0.03
	0.029	+0.075	+0.031	−0.746	.942	0.02
Thigh	*Weight (kg)*	*Upper Thigh C*	*Iliac Cr Fat**			
	0.120			−1.123	.893	0.54
	0.074	+0.138		−4.641	.933	0.45
	0.074	+0.123	+0.027	−4.216	.944	0.43
Calf	*Calf Circ*	*Tibiale Ht*	*Ankle Circ*			
	0.135			−1.318	.933	0.14
	0.141	+0.042		−3.421	.971	0.09
	0.111	+0.047	+0.074	−4.208	.979	0.08
Foot	*Weight (kg)*	*Ankle Circ*	*Foot Length*			
	0.009			+0.369	.810	0.06
	0.005	+0.033		−0.030	.882	0.05
	0.003	+0.048	+0.027	−0.869	.907	0.04

* Iliac Cr Fat (mm) = 0.78 Skinfold. Iliac Crest (mm) − 0.27; weight is kg; and all other dimensions are in cm.

employed a somewhat different method of dismembering their cadavers by using specific bony landmarks to identify the segmental endpoints. Although this system facilitates the application of their data to living subjects, it does make precise comparison of their results with those of previous investigators rather difficult. It should

Table 5-4. Segmental Weight/Body Weight Ratios from Several Cadaver Studies*

Source	Harless (1860)	Braune and Fischer (1889)	Fischer (1906)	Dempster (1955)	Dempster† (1955)	Clauser et al. (1969)
Sample Size	2	3	1	8	8	13
Head	7.6%	7.0%	8.8%	7.9%	(8.1)%	7.3
Trunk	44.2	46.1	45.2	48.6	(49.7)	50.7
Total Arm	5.7	6.2	5.4	4.9	(5.0)	4.9
Upper Arm	3.2	3.3	2.8	2.7	(2.8)	2.6
Forearm &	2.6	2.9	2.6	2.2	(2.2)	2.3
Hand	1.7	2.1	—	1.6	(1.6)	1.6
Forearm	0.9	0.8	—	0.6	(0.6)	0.7
Hand	18.4	17.2	17.6	15.7	(16.1)	16.1
Total Leg	11.9	10.7	11.0	9.7	(9.9)	10.3
Thigh	6.6	6.5	6.6	6.0	(6.1)	5.8
Calf & Foot	4.6	4.8	4.5	4.5	(4.6)	4.3
Calf	2.0	1.7	2.1	1.4	(1.4)	1.5
Foot						
Sum	100.0	100.0	100.0	97.7	100.0	100.0

*(Clauser, C. E., McConville, J. T., and Young, J. W.: Weight, Volume, and Center of Mass of Segments of the Human Body. Wright-Patterson Air Force Base, Ohio, p. 59, 1969. AMRL-TR-69-70).

†A proportional adjustment has been made to each of the segments to account for fluid and tissue losses and bring the sum of the parts to 100%.

Table 5-5. Segmental Center of Mass Locations with Respect to the Upper Boundary of the Segment

Body Segment	Harless (1860) Kefer	Harless (1860) Graf	Braune and Fischer (1889) N = 3	Dempster (1955) N = 8	Clauser et al. (1969) N = 13
Head	.36	.37			.47
Trunk			.61†		.38
Upper Torso	.50	.43		.63	
Lower Torso	.52	.44		.60	
Thigh	.43	.47	.44	.43	.37
Shank	.44	.36	.42	.43	.37
Foot*	.44	.46	.43	.44	.45
Upper Arm	.43	.49	.47	.44	.51
Forearm	.42	.44	.42	.43	.39
Hand	.36	.47		.51	.18

*From heel.

†Measured from the atlanto-occipital joint.

be noted that they found that three-step regression equations were superior to the simple ratio method of predicting the location of the mass center.

There is a dearth of cadaver data which can be used to compute segmental moments of inertia in living subjects. Rather than relating this parameter directly to limb length or mass, the general practice has been to calculate radii of gyration from the moments of inertia determined in the cadaver studies. These radii have subsequently been expressed as a proportion of segment length. Fischer (1906) stated that the radius of gyration for moments of inertia of the greater extremities with respect to all axes passing through the mass center and oriented perpendicular to the long axis of the limb was three-tenths of the segment length (Figure 5-1). This relationship also

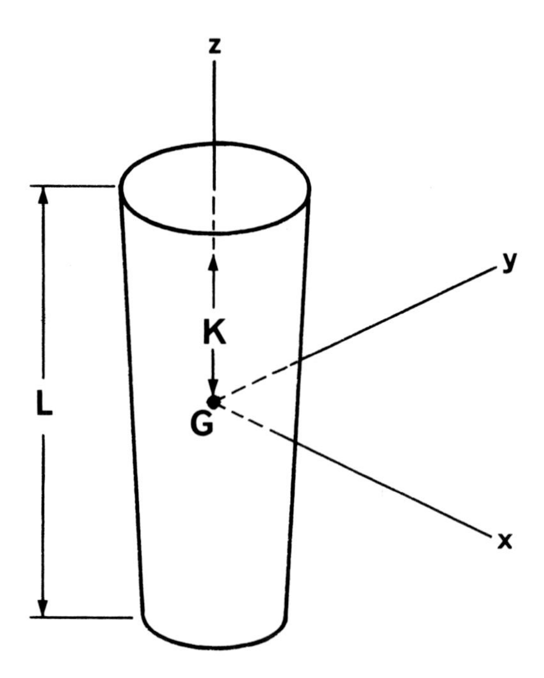

I_G Moment of Inertia with Respect to an Axis through the Mass Center
L Segment Length
k Radius of Gyration = .3 L

$$I_{G_{xx}} = mk^2 = m(.3\ L)^2$$
$$I_{G_{yy}} = mk^2 = m(.3\ L)^2$$

FIGURE 5-1. *Radius of Gyration* (*O. Fischer,* Theoretische Grundlagen für eine Mechanik der Lebenden Körper mit Speziellen Anwendungen auf den Menschen sowie auf einige Bewegungsvorgänge an Maschinen. 1906, Teubner: Leipzig).

applied to moments of inertia with respect to axes parallel to the hip axis and passing through the mass center of the head and trunk. Using Dempster's (1955a) data on moments of inertia, Plagenhoef (1971) calculated the segmental radii of gyration for moments of inertia with respect to axes perpendicular to the long axis of the limb and passing through either the proximal or the distal ends (Table 5-6). This information from Plagenhoef and Fischer, in conjunction with appropriate segmental masses and the parallel axis theorem, makes it possible to calculate the moments of inertia with respect to axes perpendicular to the length of the segment which are parallel to those passing through the mass center.

All of these computational methods, whether based upon direct proportions or regression equations, can provide information on segmental masses, mass centers and selected moments of inertia. However, the source from which they were derived, namely, a limited number of Caucasian adult male cadavers, must be kept in mind. Investigators using these methods must always be cognizant of this limitation, particularly if they intend to extrapolate to populations differing in age, sex, race or body composition from the original sample.

Immersion. The immersion or water displacement method has been employed by a number of investigators to determine the segmental masses of living subjects. This procedure, which is based upon the principle that a body segment, when immersed in water, will displace an amount of fluid equivalent to its volume, is quite well adapted to measuring the extremities but may present difficulties if used with the trunk and head. Nevertheless, Zook (1932) managed to adapt it to the whole body when he tested 164 boys ranging in age from 5 to 19 years. He lowered each subject into a

Table 5-6. Proximal and Distal Radii of Gyration*

Segment	*Proportion of Segment Length from*	
	Proximal End	*Distal End*
Head and Trunk	.830	.607
Thigh	.540	.653
Shank	.528	.643
Foot	.690	.690
Upper Arm	.542	.645
Forearm	.526	.647
Hand	.587	.577

*(Dempster, W. T.: Space Requirements of the Seated Operator. Wright-Patterson Air Force Base, Ohio, 1955 WADC TR 55-159).

tank of water maintained at body temperature. The overflow was measured as the water level reached the ankle, knee, crotch, navel, bottom of the sternum, top of the sternum, auditory canal and top of the head. A rubber tube was provided for breathing when the nose and mouth were submerged.

Dempster (1955a) used specially designed tanks to accommodate the various segments. The tanks were filled to the brim with water. The subject then lowered his limb into the water until it reached a predetermined anatomical boundary level. Water displaced by the segment was collected in an overflow catch jacket surrounding the tank. Dempster assumed that the density of all the segments was the same as that of water; one gram per cubic centimeter. Since

$$\begin{aligned} \text{mass (gm)} &= \text{density (gm/cc)} * \text{volume (cc)} \\ &= 1 * \text{volume (cc)}, \end{aligned}$$

the weight of the displaced water in grams was taken to be equivalent to the segmental mass in grams.

Clauser and his co-workers (1969) used a similar method but also included a correction for water temperature. They noted that, while repeated volume measurements of cadaver segments fell within a range of ±.5% of the segment's average volume, a variation as high as 3% to 5% was recorded when using living subjects. This increased error was attributed to subject movement during the measurement.

Further refinement of the instrumentation was provided by Drillis and Contini (1966). Their immersion technique, involving the use of two connected, uniform cylinders graduated in centimeters, appears to be the most efficient yet developed. One cylinder, which is elevated and contains the most water initially, is designated the supply tank. The other, in which the limb is placed, is referred to as the measurement cylinder (Figure 5-2). Water is allowed to flow from the supply into the measurement tank until it reaches a distal reference point marked on the segment under consideration. The level of the water in both cylinders is then recorded. The water is subsequently permitted to flow into the measurement cylinder until it reaches the proximal anatomical reference indicating the top of the segment being evaluated. Water levels are again recorded. The differences between the initial and final water levels are converted to volumes by multiplying by the cross-sectional area of the particular cylinder. The difference in the two volumes will indicate the volume of the segment. If the density of the segment is assumed to be 1.00, and the water temperature has been maintained at 4°C., then the volume in centimeters is equivalent to the mass of the segment in grams. This technique provides more accurate results than those obtained previously. The problem of water spillage, a source of error

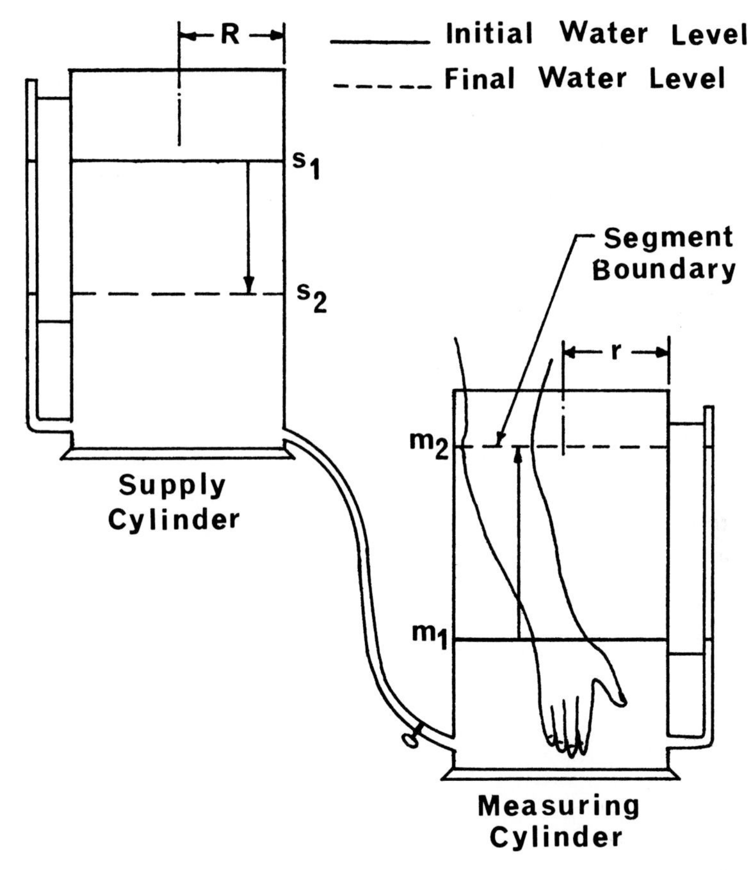

Segment Volume = Δ Measuring Cylinder Volume − Δ Supply Cylinder Volume
Segment Volume (cc) = $(m_2 - m_1) * \pi r^2 - (s_2 - s_1) * \pi R^2$
Segment Mass (gm) = Segment Volume (cc) * Segment Density (gm/cc)
= Segment Volume (cc) * 1.00

FIGURE 5-2. *Immersion Technique.*

in overflow methods, is obviated. In addition, since the subject does not have to submerge his limb, he can assume a stationary and thus more comfortable position while the water level rises during the course of the measurement.

Calculation of segmental mass distribution based upon the immersion principle requires a knowledge of the number of cubic centimeters of water displaced and the density of the limb. If a segment density of one gram per cubic centimeter is not considered sufficiently accurate for the purposes of the study, then the values determined by Harless (1860) or Dempster (1955a) (Table 5-7) could be inserted into the formula.

Table 5-7. Body Segment Densities from Cadaver Studies (Grams/Cubic Centimeter)

Segment	*Harless** *(1860)*	*Dempster†* *(1955)*
Head and Neck	1.11	1.11
Trunk	—	1.03
Upper Arm	1.08	1.07
Forearm	1.10	1.13
Hand	1.11	1.16
Thigh	1.07	1.05
Lower Leg	1.10	1.09
Foot	1.09	1.10

* Based upon the dissection of five cadavers.
† Based upon the dissection of eight cadavers.

The immersion method can also be utilized to locate the segmental mass centers in living subjects. Cleaveland (1955) employed a large tank filled with water into which he could lower a subject in a hammock or on a platform. The weight of the subject in the air was indicated on a balance scale from which he and the hammock were suspended. He was then lowered so that a body segment was immersed in the water. The weight (immersion weight) was again recorded. According to Archimedes' principle, the apparent weight loss on immersion equals the weight of the water displaced. This coincides with the volume of the segment. Using the calculation,

$$\frac{\text{air weight} - \text{immersion weight}}{2} + \text{immersion weight}$$

and assuming the mass center to be coincident with the center of volume, Cleaveland determined what he called the "center of gravity weight." The limb was then withdrawn until the scale registered this particular value. The water level on the limb at that weight indicated the location of the center of gravity. A simpler version of this technique employs the two connected cylinders described earlier. The segmental volume is determined by allowing water to flow into the measurement cylinder. The water level is then dropped until half of the volume is back in the supply tank.

A segmental zone approach can also be taken to the immersion technique. By successively displacing small volumes (v_i) of water and measuring the distance ($\mathbf{r}_i$) from the end of the limb to the center of each volume element, it is possible to compute the location of the center of mass. The mass moment of inertia about any axis perpendicular to the long axis of the limb can also be calculated. These concepts are illustrated in Figure 5-3.

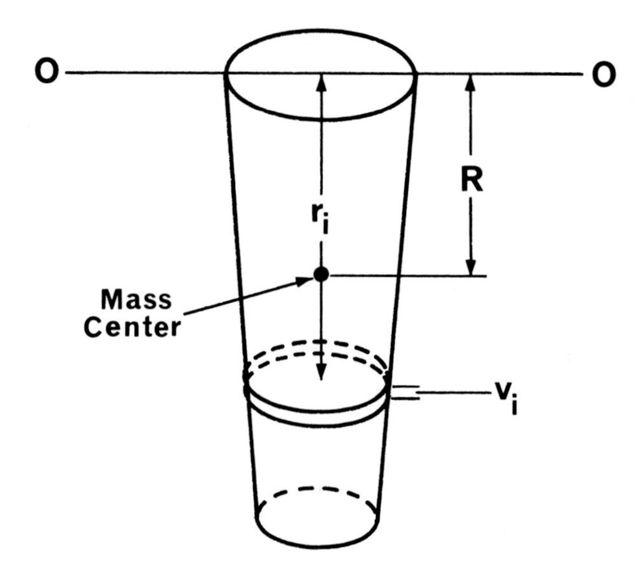

Mass = Density * Volume
$\Sigma m_i = \rho v_i$

Location of the mass center with respect to the 00 axis:

$$RM = \sum_{i=1}^{N} m_i r_i$$

$$R = \frac{\sum_{i-1}^{N} m_i r_i}{M}$$

in which $M = \sum_{i=1}^{N} m_i$ = total mass of the segment

Moment of inertia with respect to the 00 axis:

$$I_{oo} = \sum_{i=1}^{N} m_i r_i^2$$

FIGURE 5-3. *Segmental Zone Approach to the Immersion Technique.*

Reaction Change. The reaction change or board and scale technique for estimating segmental weight has been described by Williams and Lissner (1962) and Drillis and Contini (1966). There is also some evidence to suggest that the Russian physiologist, Bernstein, utilized it in his study of 152 living subjects in the 1930's. Like immersion, the reaction-change technique is best suited to the determination of the weights of the extremities rather than of the head and trunk.

The procedure and calculations required (Williams and Lissner, 1962) are outlined in Figure 5-4. The subject lies either prone or supine on a horizontal board which is fixed at one end and supported at the other by a scale or other type of force transducer. A measuring tape is affixed to the side of the board. In the diagram, B indicates the weight of the board; W, the subject's weight; Sy, the scale reading; $W1$, the weight of the subject minus the segment for which the weight is to be calculated; and $W2$ is the weight of the segment under consideration. All weights are shown acting at their respective centers of gravity. Since the system (board and subject) is in equilibrium, the sum of the moments about any axis is equal to zero. To eliminate unknown reaction forces from the calculation, moments are taken about an axis through R as shown in equation (1). The perpendicular distances from the lines of action of the forces to the axis are indicated by the horizontal lines. In accordance with the principle of moments which states that the sum of the moments is equal to the moment of the sum, relationships (2) and (4) are derived. In both cases, the body weight is as follows: $W = W1 + W2$. After assuming the initial position shown in Figure 5-4A and 5-4B, the arm is raised to a position perpendicular to the board (Figure 5-4C and 5-4D). Although any other position which differs from the original would be theoretically acceptable, the perpendicular limb orientation is maintained more easily by the subject while the necessary measurements are being recorded.

Calculations make evident that only these five variables must be recorded in order to determine segmental weight: l, $x2$, and $xx2$, horizontal distance between the axis through R and the scale, line of action of $W2$ with the segment in the initial position, and line of action of $W2$ with the segment in the perpendicular position, respectively; and Sy and SSy, the scale readings with the body in the initial prone or supine position and with the designated limb in the perpendicular position. One of the chief limitations of this technique is that center of mass of the segment must be known before segmental weight can be found. If it is not located correctly, further error will be introduced.

The reaction change method cannot be used to estimate segmental moments of inertia and has only limited application in locating segmental mass centers. In the latter case, the same procedure can be used for the segment as for the total body† provided that a representative cast of the limb is placed in a known position on the board.

†Refer to Chapter 3, pages 21–23.

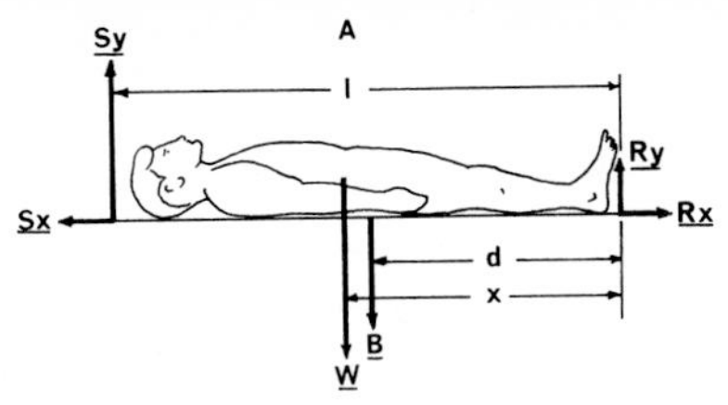

$$\Sigma M_R = 0$$

$$Sy * l - W * x - B * d = 0$$

$$x = \frac{Sy * l - B * d}{W} \quad (1)$$

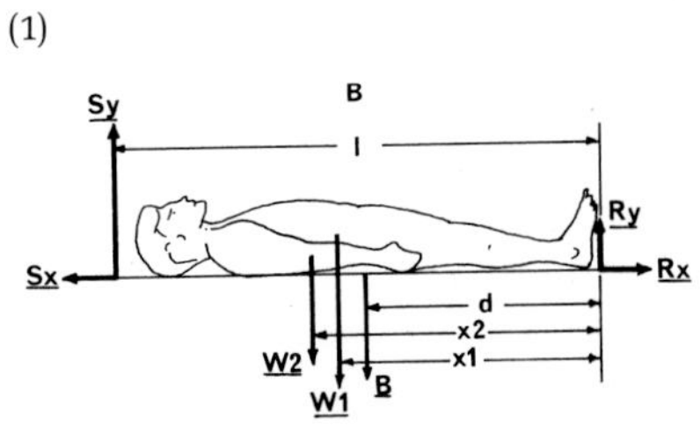

Principle of Moments

$$W * x = W1 * x1 + W2 * x2 \quad (2)$$

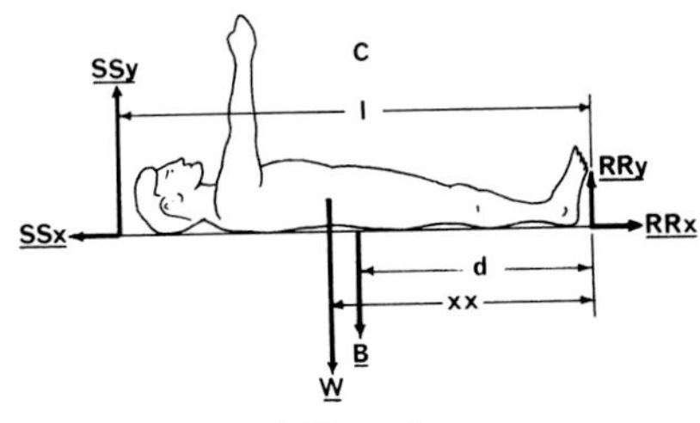

$$\Sigma M_R = 0$$

$$SSy * l - W * xx - B * d = 0$$

$$xx = \frac{SSy * l - B * d}{W} \quad (3)$$

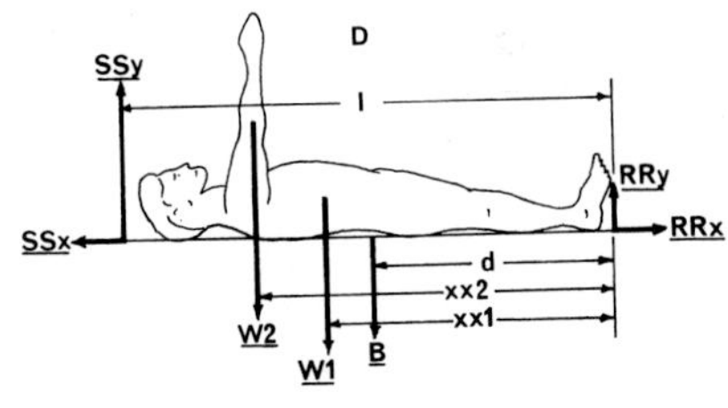

Principle of Moments

$$W * xx = W1 * x1 + W2 * xx2 \quad (4)$$

Calculation of Segmental Mass

Subtracting (4) from (2)

$$\begin{aligned} W * x &= W1 * x1 + W2 * x2 \\ W * xx &= W1 * x1 + W2 * xx2 \\ \hline W(x - xx) &= W2(x2 - xx2) \end{aligned}$$

$$W2 = \frac{W(x - xx)}{(x2 - xx2)} \quad (5)$$

Substituting (1) and (3) into (5)

$$W2 = \frac{W\left(\frac{Sy * l - B * d}{W} - \frac{SSy * l - B * d}{W}\right)}{(x2 - xx2)}$$

$$W2 = \frac{l(Sy - SSy)}{(x2 - xx2)}$$

FIGURE 5-4. *Determining Segmental Weight Utilizing the Method of Reaction Change (Adapted from Williams, M., and Lissner, H. R.: Biomechanics of Human Motion. Philadelphia: W. B. Saunders, 1962).*

Photogrammetry. The photogrammetric technique of determining body surface area has been described by several investigators (Berner, 1954; Hertzberg, 1957; Pierson, 1957, 1959, 1961a, 1961b, 1963; Drillis and Contini, 1966; and Herron, 1969, 1970) but has not been used widely to collect body segment parameter data. Most writers have emphasized the total body rather than the individual segment application of mono-and stereo-photogrammetry. The method itself is based upon the same principles employed in constructing contour maps from aerial photographs. Contours must first be projected onto the subject. Pierson (1963) suggested that this be accomplished by illuminating the subject on both sides by lights shining through colored transparent stripes of equal width. The subject could then be photographed from the back and the front with the projected colors indicating varying levels of the body surface. The areas occupied between the contours are subsequently calculated and volumes determined utilizing the appropriate photogrammetric procedures. The technique requires very little of the subject's time and thus has the advantage of rapid data collection. The data reduction, a very laborious task if performed by hand, has been facilitated by the use of automatic plotting devices and computer processing.

Radiation Techniques. Perhaps the most promising new approach to the determination of body segment parameters on living subjects was reported by Casper and his associates (1971). Their method involved the focusing of high-energy gamma rays (Cobalt 60) onto a selected object. Since the transmission of the photons is related to the mass and relatively independent of the elemental composition of the material traversed, the number of rays penetrating the object provides a basis for evaluating its mass. The prototype used by Casper and his co-workers to determine the mass, center of mass and moments of inertia of metal, wood and plastic test objects is presented in Figure 5-5. The system, shown in block diagram form (Figure 5-6), includes the scan table, detector assembly, data amplification and analysis, digital computer and teletype for the print-out of results. The digital computer controls the scanning operation which is performed at one-second intervals over each one-fourth inch along the X axis. Data from the counter are accepted by the computer, stored and later used to calculate the desired parameters. Results from this gamma mass scanner were within ±1% of the criteria, indicating that the system was indeed capable of accurate measurement.

A subsequent validation study by Brooks (1973) utilized biological tissue in the form of legs of lamb (Figure 5-7). The scanner estimates of the body segment parameters were compared with those derived

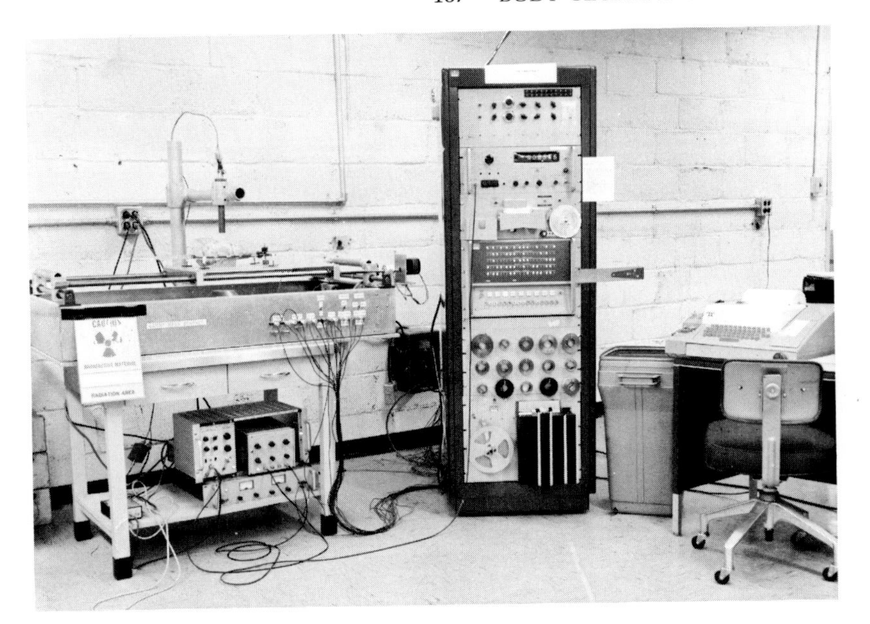

FIGURE 5-5. *Gamma Mass Scanner.*

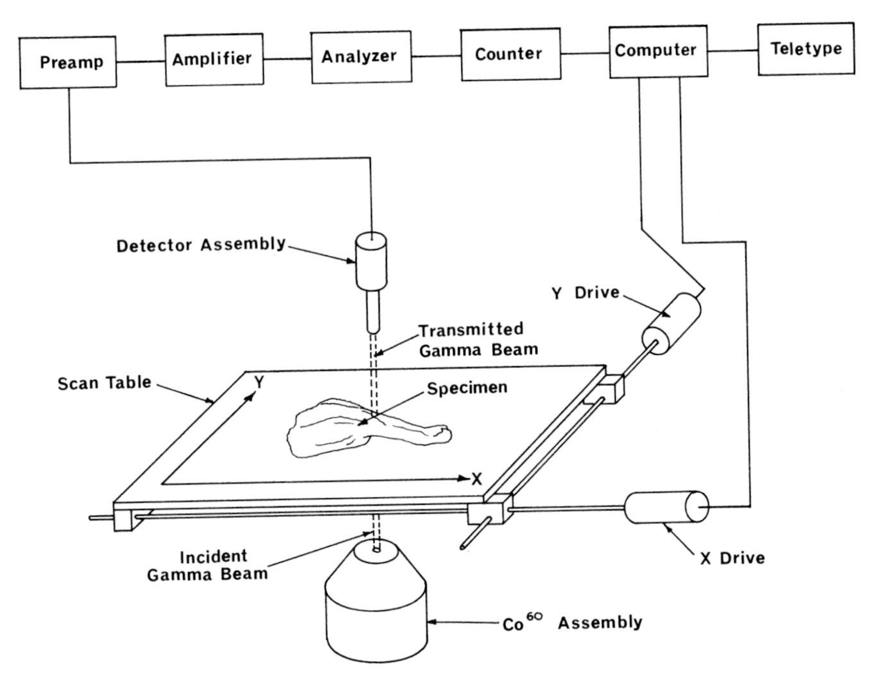

FIGURE 5-6. *Block Diagram of Gamma Mass Scanner.*

FIGURE 5-7. *Leg of Lamb Positioned on Scanner Table.*

from immersion, reaction change and pendulum methods. Although the presence of hydrogen affected the gamma ray scattering, the results indicated that this was not a serious problem and that accurate values could be obtained.

These preliminary investigations have shown that the principle of using gamma radiation to determine segmental mass, center of mass and moments of inertia is valid. Development of a model with brief scanning time and low radiation levels is now possible. Such a system may help to solve the perplexing problem of obtaining body segment parameters in living subjects.

Other Methods. In addition to the techniques already described for determining the segmental parameters of the human body, there are other methods which have been cited in the literature. In some cases, a plaster cast has been made of a limb and its mass determined by immersion; its mass center by a balancing or reaction change technique; and its moment of inertia with respect to a horizontal axis by suspending the cast and swinging it as a pendulum (Drillis and Contini, 1966). An obvious question can be raised as to the similarity between the body segment and the plaster cast in terms of mass distribution and density.

The quick release technique (Drillis and Contini, 1966) has been proposed as a means of estimating the moment of inertia of a segment with respect to a horizontal axis passing through a proximal joint. It is best suited for the forearm-hand and shank-foot segments. The limb is positioned in a vertical plane and maintains tension in a cord which is attached to a force transducer. Two linear accelerometers, a known distance apart, are fastened to a limb to record the tangential component of acceleration. They are wired to provide a differential output which neutralizes the effect of gravity and gives the angular acceleration of the segment being measured. Since the sum of the moments of force about a fixed axis is equal to the moment of inertia about the axis multiplied by the angular acceleration, the quick release method can be used to determine that moment of inertia (Figure 5-8).

Despite the existence of a large number of experimental techniques for estimating the body segment parameters of living subjects, many problems remain. It is difficult to obtain data on the head and trunk. No satisfactory method has yet been employed to determine the three principal moments of inertia of the segments. The basic data used in the computational methods are not intended for use with women and children. In addition, these methods would not be administratively feasible for a study involving a large number of subjects in which masses, mass centers and moments of inertia were required for all the segments. Such information would be necessary for a complete kinetic analysis of a sports skill.

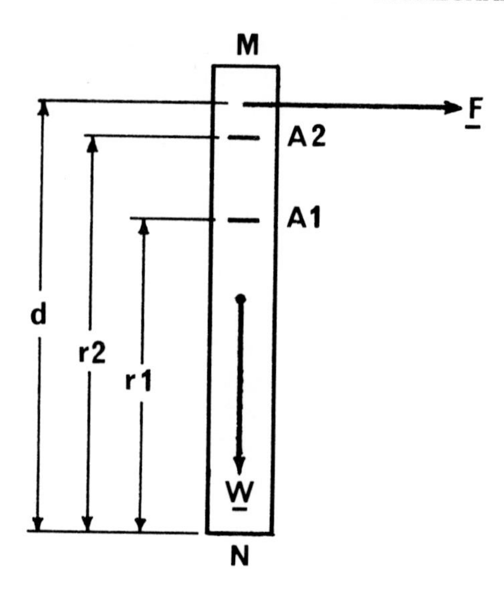

MN	Body segment
F	Force exerted by the force transducer
W	Segment weight
A1, A2	Accelerometers one and two
r1, r2	Distance from the axis of rotation through N to accelerometers *A1* and *A2* respectively
α	Angular acceleration of the segment
a_{t_1}, a_{t_2}	Tangential components of acceleration of the segment at *A1* and *A2* respectively
d	Distance from the axis of rotation through N to the force transducer

Calculation of Moment of Inertia about a Horizontal Axis through N

$$a_{t_2} = r2\ \alpha = (r2 - r1 + r1)\ \alpha$$

$$a_{t_1} = r1\ \alpha$$

$$a_{t_2} - a_{t_1} = (r2 - r1 + r1 - r1)\ \alpha = (r2 - r1)\ \alpha$$

$$\alpha = \frac{\text{Differential accelerometer output}}{\text{Distance between accelerometers}}$$

$$\Sigma M_N = I_N\ \alpha$$

$$F * d = I_N\ \alpha$$

$$I_N = \frac{F * d}{\alpha}$$

FIGURE 5-8. *Quick Release Method for Determining Moment of Inertia.*

MATHEMATICAL MODELS

With the increasingly widespread use of the digital computer in biomechanical research, it has become possible to carry out extensive calculations of body segment parameters quickly and accurately.

The capabilities of the computer have thus encouraged the development of mathematical models of the body. Such models characteristically employ a number of simplifying assumptions concerning the composition and functioning of the body. Joints are usually considered to be frictionless and pinned. Displacement of blood and soft tissue is disregarded as segments are assumed to be homogeneous rigid bodies of simple geometrical shape. Hands and feet are described as single segments. Data from the dissection studies of Braune and Fischer (1889), Fischer (1906) and Dempster (1955a) have provided segmental mass proportions. Body weight in conjunction with a series of anthropometric measurements of the subject has furnished the necessary information to define segment lengths, circumferences and masses. Because of the simplified geometrical configurations of the idealized body parts, determinations of the mass center locations and the principal moments of inertia have become straightforward mathematical operations.

One of the earlier attempts to describe the body in mathematical terms was made by Kulwicki and Schlei (1962). Their model consisted of six right circular cylinders representing the arms, legs, head and torso. Segmental parameters were obtained from Dempster's study (1955a), fiftieth percentile data from the 1950 survey of USAF personnel and information from the Anthropology Section of the Aerospace Medical Research Laboratories at Wright-Patterson Air Force Base. The model, however, was not verified for accuracy nor was it adapted to suit men of differing proportions.

Whitsett (1963), in a study of the dynamic characteristics of men under zero gravity conditions, formulated a mathematical model of the body composed of 14 rigid, homogeneous segments having simple geometrical configurations. The model was considered hinged at the joints and possessed 24 degrees of freedom. Barter's regression equations (1957) estimated the masses of all the segments with the exception of the head and torso while Dempster's data (1955a) were employed for the average segmental densities as well as for the location of the mass centers of the upper and lower arms and legs. The centers of mass of the other segments were situated halfway along their axes of symmetry. Anthropometric measurements of living subjects specified the segmental dimensions. Whitsett determined the relative contributions of local and transfer terms in the moments of inertia of the segments about axes through the mass center of the body and found the local terms to be negligible for the hands, feet and forearms. Upon this basis, he constructed a simplified system for determining the total body moments of inertia which neglected changes in the local terms of the six segments mentioned and included only their transfer terms. No experimental validation was given for the model.

A computerized 15-segment model of the body was devised by Hanavan (1964). The head was depicted as an ellipsoid of revolution, the upper and lower torso as right elliptical cylinders and the hands as solid spheres (Figure 5-9). All the other segments were portrayed as frusta of right circular cones. Twenty-five standard anthropometric measurements of the individual subjects defined the necessary lengths and diameters. Barter's regression equations (1957) were employed to approximate the mass of the segments. Hanavan utilized the information on masses, dimensions and geometrical properties of the idealized segments to calculate their principal moments of inertia and the location of their centers of mass. The model was personalized to the extent that the dimensions of any subject could be used as input to the program and the corresponding body segment parameters computed. Since Hanavan's main objective was to determine the inertial properties of the total body in several different positions, his computer program did not output data on the individual segments. This information, however, can be obtained by using the subroutine which calculates these parameters and inserting appropriate input/output statements. Other modifications are also possible.

To validate his model, Hanavan compared the computer results with those obtained experimentally on 66 subjects (Santschi *et al.*, 1963).He concluded that, in general, the total body center of gravity location was predicted within 0.7 inches of the experimental values, with especially good correspondence being noted between the vertical coordinates. In the moments of inertia, the greatest discrepancies appeared in the results for the hands and feet. These, however, did not have a serious effect upon the calculations since their moments were extremely small when compared with those of the other segments. In addition, for both hands and feet, the transfer terms in the parallel axis theorem assume much greater significance than do the local terms. On the average, the moments of inertia were predicted within 10% of the criterion. The principles of the Hanavan model have had wide research application as evidenced by their inclusion in studies by McCrank and Seger (1964), Smith and Kane (1967), Riddle and Kane (1968), Scher and Kane (1969) and Miller (1970).

Tieber and Lindemuth (1965) incorporated several modifications into the Hanavan model which they claimed improved its predictive accuracy. They used a new set of regression equations for estimating segmental weight (Table 5-8). These revisions increased the input requirements by the addition of age and two more anthropometric measurements to the original 25 stipulated by Hanavan.

Computerized mathematical representations of the total body are

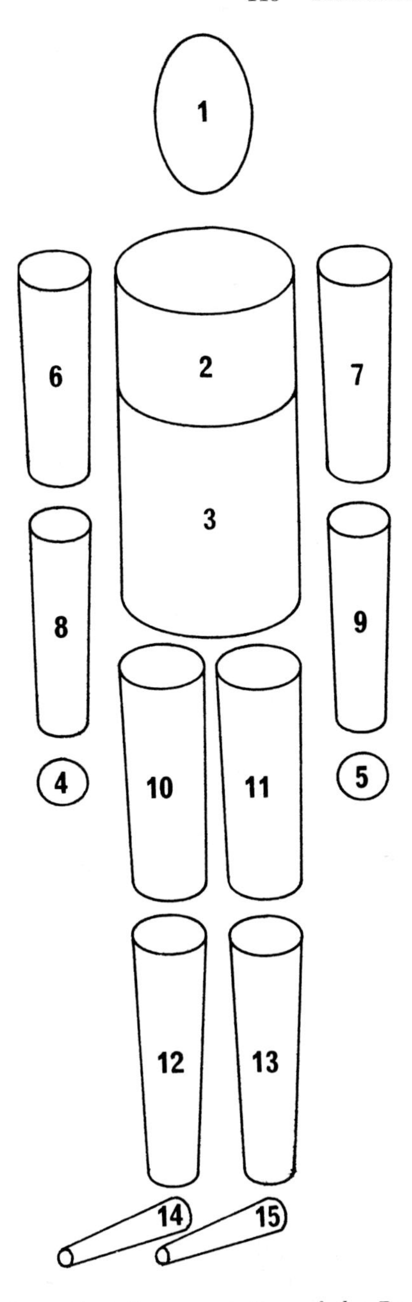

FIGURE 5-9. *Mathematical Representation of the Body (Hanavan, E. P.: A Mathematical Model of the Human Body. Wright-Patterson Air Force Base, Ohio, p. 8, 1964. AMRL-TR-64-102).*

Table 5-8. Segmental Weight Regression Equations*

Segment	*Segment Weight (lb)*
Head	.028 Body Weight (W) + 6.354
Trunk	.552 W – 6.417
Upper Arms	.059 W + .862
Forearms	.026 W + .85
Hands	.009 W + .53
Upper Legs	.239 W – 4.844
Lower Legs	.067 W + 2.846
Feet	.016 W + 1.826

*(Tieber, J. A., and Lindemuth, R. W.: An Analysis of the Inertial Properties and Performance of the Astronaut Maneuvering System. Wright-Patterson Air Force Base, Ohio, p. 10, p. 14, 1965. AMRL-TR-65-216). Information in this table is based upon unpublished data from a study by C. Clauser and J. T. McConville to determine selected characteristics of segments of the human body, Aerospace Medical Research Laboratories, Wright-Patterson Air Force Base, 1964–65.

extremely useful for studies concerned with the kinetic analysis of sport. They have the advantage of being able to define the biomechanical properties of all the segments on the basis of a few readily obtainable anthropometric measurements of the subject. They also provide a rapid and accurate means of executing the necessary calculations. This technique of estimating body segment parameters is administratively feasible for large numbers of subjects and can provide data concerning moments of inertia not available from other sources. At present, however, their scope of inference is limited to adult male subjects. The fact that the mass distribution relationships have been derived from a relatively small sample of cadavers must be kept in mind. The recent study by Clauser and associates (1969) has supplied further information on these segmental mass proportions and on center of mass locations. It is anticipated that future models will incorporate these data as well as other information of this nature as it becomes available from subsequent research.

SELECTED REFERENCES

Barter, J. T.: Estimation of the Mass of Body Segments. Wright-Patterson Air Force Base, Ohio, 1957 (WADC TR 57-260).

Berner, J.: Calculation of the Body Surface Area by Photogrammetry. Scand. J. Clin. Lab. Invest., 6, 322–324, 1954.

Bernstein, N. A.: *The Co-ordination and Regulation of Movements.* New York: Pergamon Press, 1967.

Boyd, E.: The Specific Gravity of the Human Body. Hum. Biol., *5*, 646–672, 1933.

Braune, W., and Fischer, O.: Über den Schwerpunkt des menschlichen Korpers mit Rucksicht auf die Ausrustung des deutschen Infanteristen. Abh. d. math.-phys. cl. d. K. Sachs. Gesselsch. der Wiss., *26*, 561–672, 1889. In W. M. Korgman and F. E. Johnston (Eds.), Human Mechanics—Four Monographs Abridged. Wright-Patterson Air Force Base, Ohio, 1963 (AMRL-TDR-63-123).

Brooks, C. M.: Validation of the Gamma Mass Scanner for Determination of Center of Gravity and Moment of Inertia of Biological Tissue. Unpublished Master's Thesis, Pennsylvania State University, 1973.

Casper, R. M.: On the Use of Gamma Ray Images for Determination of Human Body Segment Parameters. Unpublished Master's Thesis, Pennsylvania State University, 1971.

Casper, R. M., Jacobs, A. M., Kenney, E. S., and McMaster, I. B.: On the Use of Gamma Ray Images for Determination of Human Body Segment Parameters. Paper presented at Quantitative Imagery in Biomedical Sciences, Houston, Texas, 1971.

Clauser, C. E., *et al.*: Anthropometry of Air Force Women. Wright-Patterson Air Force Base, Ohio, 1972 (AMRL-TDR-72-5).

Clauser, C. E., McConville, J. T., and Young, J. W.: Weight, Volume, and Center of Mass of Segments of the Human Body. Wright-Patterson Air Force Base, Ohio, 1969 (AMRL-TR-69-70).

Cleaveland, H. G.: The Determination of the Center of Gravity of Segments of the Human Body. Unpublished Doctoral Dissertation, University of California, Los Angeles, 1955.

Contini, R., Drillis, R., and Bluestein, M.: Determination of Body Segment Parameters. Hum. Factors, *5*, 493–504, 1963.

Damon, A., Stoudt, H. W., and McFarland, R. A.: *The Human Body in Equipment Design*. Cambridge, Mass.: Harvard University Press, 1966.

Dempster, W. T.: Free Body Diagrams as an Approach to the Mechanics of Human Posture and Motion. In F. G. Evans (Ed.), *Biomechanical Studies of the Musculo-Skeletal System*. Springfield, Ill.: C. C Thomas, 1961.

Dempster, W. T.: Space Requirements of the Seated Operator. Wright-Patterson Air Force Base, Ohio, 1955a (WADC TR 55-159).

Dempster, W. T.: The Anthropometry of Body Action. Ann. N.Y. Acad. Sci., *63*, 559–585, 1955b.

Dempster, W. T., Sherr, L. A., and Priest, J. G.: Conversion Scales for Estimating Humeral and Femoral Lengths and the Lengths of Functional Segments in the Limbs of American Caucasoid Males. Hum. Biol., *36*, 246–262, 1964.

Drillis, R., and Contini, R.: Body Segment Parameters. School of Engineering and Science, New York University, 1966 (PB 174 945; Tech. Rpt. No. 1166.03).

Drillis, R., Contini, R., and Bluestein, M.: Body Segment Parameters—A Survey of Measurement Techniques. Artif. Limbs, *8*, 44–66, 1964.

Duggar, B. C.: The Center of Gravity of the Human Body. Hum. Factors, *4*, 131–148, 1962.

Fischer, O.: *Theoretische Grundlagen für eine Mechanik der Lebenden Körper mit Speziellen Anwendungen auf den Menschen sowie auf einige Bewegungsvorgänge an Maschinen.* Leipzig: Teubner, 1906.

Fujikawa, K.: The Center of Gravity in the Parts of the Human Body. Okijimas Folia Anat. Jap., *39*, 117–125, 1963.

Geohegan, B.: The Determination of Body Measurements, Surface Area and Body Volume by Photography. Amer. J. Phys. Anthropol., *11*, 97–119, 1953.

Goto, K., and Shikko, H.: On the Measurement of Weight and Position of Center of Gravity of Parts of Body. Tokyo Izishinshi, *73*, 1956 (English translation by S. Katoh).

Hanavan, E. P.: A Mathematical Model of the Human Body. Wright-Patterson Air Force Base, Ohio, 1964 (AMRL-TR-64-102).

Harless, E.: The Static Moments of the Component Masses of the Human Body. 1860, Wright-Patterson Air Force Base, Ohio, 1962 (FTD-TT-61-295).

Hay, J. G.: The Center of Gravity of the Human Body. In *Kinesiology 1973.* Washington: AAHPER, 1973.

Herron, R. E.: Stereophotogrammetry in Biology and Medicine. Photogr. Appl. Sci. Technol. Med., *5*, 26–35, 1970 (September).

Herron, R. E.: A Biomedical Perspective in Stereographic Anthropometry. In F. D. Thomas and E. Sellers (Eds.), *Biomedical Instrumentation.* Vol. 6. Pittsburgh: Instrument Society of America, 1969.

Hertzberg, H. T. E., Dupertius, C. W., and Emanuel, I.: Stereophotogrammetry as an Anthropometric Tool. Photogram. Eng., *23*, 942–947, 1957.

Krogman, W. M., and Johnston, F. E. (Eds.): Human Mechanics—Four Monographs Abridged. Wright-Patterson Air Force Base, Ohio, 1963 (AMRL-TDR-63-123).

Kulwicki, P. V., and Schlei, E. J.: Weightless Man: Self-Rotation Techniques. Wright-Patterson Air Force Base, Ohio, 1962 (AMRL-TDR-62-129).

Liu, Y. K., Laborde, J. M., and Van Buskirk, W. C.: Inertial Properties of a Segmented Cadaver Trunk: Their Implications in Acceleration Injuries. Aerospace Med., *42*, 650–657, 1971.

Matsui, H.: *A New Method to Determine the Center of Gravity of a Human Body by Somatometry.* Tokyo: Taiiku no Kagakusha, 1958 (Japanese).

McCrank, J. M., and Seger, D. R.: Torque Free Rotational Dynamics of a Variable-Configuration Body (Application to Weightless Man). Wright-Patterson Air Force Base, Ohio 1964 (GAW/Mech 64-19).

McHenry, R. R., and Naab, K. N.: Computer Simulation of the Crash Victim—A Validation Study. Proceedings of the 10th Stapp Car Crash Conference, Hollman Air Force Base, New Mexico, 1966.

Meeh, C.: Volummessungen des menschlichen Korpers und seiner einzelnen Theile in den verschiedenen Altersstufen. Z. Biol., *31*, 125–147, 1895.

Miller, D. I.: A Computer Simulation Model of the Airborne Phase of Diving. Unpublished Doctoral Dissertation, Pennsylvania State University, 1970.

Mori, M., and Yamamoto, T.: Die Massenanteile der einzelnen Korperabschnitte der Japaner. Acta Anat., *37*, 385–388, 1959.

Parks, J. L.: An Electromyographic and Mechanical Analysis of Selected Abdominal Exercises. Unpublished Doctoral Dissertation, University of Michigan, 1959.

Pierson, W. R.: A Photogrammetric Technique for the Estimation of Surface Area and Volume. Ann. N.Y. Acad. Sci., *110*, 109–112, 1963.

Pierson, W. R.: Monophotogrammetric Determination of Body Volume. Ergonomics, *4*, 213–217, 1961a.

Pierson, W. R.: Photogrammetric Determination of Surface Area. Photogram. Eng., *27*, 99–102, 1961b.

Pierson, W. R.: The Validity for Stereophotogrammetry in Volume Determination. Photogram. Eng., *25*, 83–85, 1959.

Pierson, W. R.: Non-Topographic Photogrammetry as a Research Technique. FIEP Bull., *27*, 48–49, 1957.

Plagenhoef, S.: *Patterns of Human Motion—A Cinematographic Analysis.* Englewood Cliffs, N.J.: Prentice-Hall, 1971.

Plagenhoef, S. C.: Methods for Obtaining Kinetic Data to Analyze Human Motions. Res. Q. Amer. Assoc. Health Phys. Ed., *37*, 103–112, 1966.

Raymond, P. R.: Techniques for Measurement of Centers of Gravity. Master's Thesis, Texas A & M University, 1971 (AD-739 439).

Riddle, C., and Kane, T. R.: Reorientation of the Human Body by Means of Arm Motions. Division of Engineering Mechanics, Stanford University, 1968 (Tech. Rpt. No. 182).

Santschi, W. R., DuBois, J., and Omoto, C.: Moments of Inertia and Centers of Gravity of the Living Human Body. Wright-Patterson Air Force Base, Ohio, 1963 (AMRL-TDR-63-36).

Scher, M. P., and Kane, T. R.: Alteration of the State of Motion of a Human Being in Free Fall. Division of Applied Mechanics, Stanford University, 1969 (Tech. Rpt. No. 198).

Smith, P. G., and Kane, T. R.: The Reorientation of a Human Being in Free Fall. Division of Engineering Mechanics, Stanford University, 1967 (Tech. Rpt. No. 171).

Spivak, C. D.: Methods of Weighing Parts of the Living Human Body. JAMA *65*, 1707–1708, 1915.

Swearingen, J. J.: Determination of Centers of Gravity of Man. Federal Aviation Administration, Oklahoma, 1962 (ASTIA AD-10 410).

Swearingen, J. J., Badgley, J. M., Braden, G. E., and Wallace, T. F.: Determination of Centers of Gravity of Infants. Federal Aviation Administration, Oklahoma, 1969 (AM 69-22).

Swearingen, J. J. and Young, J. W.: Determination of Centers of Gravity of Children, Sitting and Standing. Federal Aviation Administration, Oklahoma, 1965 (AM 65-23).

Tieber, J. A., and Lindemuth, R. W.: An Analysis of the Inertial Properties and Performance of the Astronaut Maneuvering System. Wright-Patterson Air Force Base, Ohio, 1965 (AMRL-TR-65-216).

Walton, J. S.: A Template for Locating Segmental Centers of Gravity. Res. Q. Amer. Assoc. Health Phys. Ed., *41*, 615–618, 1970.

Waterland, J. C., and Shambes, G. M.: Biplane Center of Gravity Procedures. Percept. Motor Skills, *30*, 511–514, 1970.

Weinbach, A. P.: Contour Maps, Center of Gravity, Moment of Inertia and Surface Area of the Human Body. Hum. Biol., *10*, 356–371, 1938.

Weissman, S.: Anthropometric Photogrammetry. Photogram. Eng., *34*, 1134–1140, 1968.

Whitsett, C. E.: Some Dynamic Response Characteristics of Weightless Man. Wright-Patterson Air Force Base, Ohio, 1963 (AMRL-TDR 63-18).

Williams, M., and Lissner, H. R.: *Biomechanics of Human Motion*. Philadelphia: W. B. Saunders, 1962.

Zook, D. E.: The Physical Growth of Boys, A Study by Means of Water Displacement. Amer. J. Dis. Children, *43*, 1347–1432, 1932.

CHAPTER 6

Photoinstrumentation

PHOTOINSTRUMENTATION encompasses a wide range of optical recording devices, many of which have been applied to study human motion on either a qualitative or quantitative level. The filming of basketball and football games has been common practice for several years and increasing use is now being made of photography in sports such as aquatics, gymnastics, track and field and golf. The subjective analysis of such films has greatly assisted coaches in evaluating the performance of their athletes.

Recently, videotape has begun to replace conventional motion pictures for teaching and coaching purposes. Since videotape is erasable, reusable and does not require any developing, it is more economical than film. The relatively inexpensive portable recorders are simple to operate and permit immediate playback. Thus, videotape has significant potential for instruction. Pictures taken of students performing motor skills can provide them with further insight into their own actions, a greater appreciation of the mechanics of sports skills and increased interest in improving their performance. Although it is valuable for qualitative evaluations, videotape is not generally suitable for research of a quantitative nature. The image on the video recorder is formed by an electron beam impinging upon a fluorescent screen. The beam scans the face of the cathode ray tube at a rate equivalent to 30 frames per second. While the resulting image on the videotape is acceptable for viewing purposes, it is not of sufficient resolution and linearity to permit fine measurement. In addition, the image tends to deteriorate if the tapes are stored for extended periods of time.

Cinematography has been used more frequently than any other method to examine the external mechanics of human motion from a quantitative standpoint. If certain fundamentals are observed, accurate measurements can be obtained from films of subjects performing under either competitive or controlled laboratory conditions. Since the movements of the subject are not hampered by extraneous wires, as might be the case with electronic instrumentation, they are more likely to be unaffected by the experimental protocol and thus be more representative of "normal" performance. In the past, however, one of the major limitations of cinematography was the extensive amount of time required to obtain the data from the film and to complete the data analysis. This disadvantage has now been partially offset by modern motion analyzers, automated data acquisition systems and the use of high-speed digital computers for data processing.

Before using photoinstrumentation to examine a problem related to the biomechanics of sport, the researcher should give consideration to the various types of equipment available, the procedures to be followed in filming and the methods for subsequent data reduction. Thus, a general understanding of cameras, lenses, film, calibration methods, filming fundamentals and techniques for quantitative film analysis will help the researcher to take the fullest advantage of the potential of cinematography for studying human motion.†

CAMERAS

All cameras are constructed so that they carefully control the amount of light reaching the film. When the shutter is open, light is permitted to enter through an aperture in the diaphragm and is focused onto the film by the lens. When the shutter is closed, the camera is a light-tight case. These basic principles apply to both still and motion picture cameras. In the latter, a series of single pictures are taken and then projected back at a rate which gives the illusion of motion.

† The reader's attention is directed to the book, *Focal Encyclopedia of Photography* (1969) as an extremely useful reference for photoinstrumentation.

Ciné or motion picture cameras can be classified according to the type of film which they employ. Thus, they range from 8mm (actually 16mm film which is run through the camera twice and then split into two 8mm widths after processing) and Super 8 (with smaller perforations and a larger image than 8mm) through 16mm and 35mm to 70mm. Super 8 and 8mm, while the least expensive of those mentioned, do not produce a sufficiently large image for most quantitative analyses. Because 16mm cameras provide a reasonable balance between economy and image size, they are employed almost exclusively in high-speed photography in which precise measurements must be made from film. Although 35mm and 70mm permit larger images, these cameras are not widely used for research since they are more expensive and the cost of their film and film processing is also higher. In addition, they require motion analyzers which differ from the conventional models in order to accommodate their larger film dimensions.

Still cameras also have a role to play in biomechanics research. The 35mm has been used for multiple image photography (Jones *et al.*, 1958; Nelson *et al.*, 1969) (Figure 6-1). With the shutter held open for a time exposure, strobe lighting can be employed to illuminate the subject at designated intervals or tiny lights fixed to important body landmarks can be flashed at a predetermined rate. Since this

FIGURE 6-1. *Stroboscopic-Photographic Record of an Elbow Flexion Movement.*

technique results in several images of the subject on a single frame, it is not suitable for studying overlapping types of motion. Polaroid cameras should also be considered as they have the distinct advantage of providing the finished picture within 15 seconds after the film is exposed. The Graph Check Sequence Camera is particularly appropriate for sports because it takes eight separate pictures during a selected time interval of one to 10 seconds. Although not appropriate for accurate quantitative analysis, these Polaroid records may prove useful during the initial stages of the investigation by providing an indication of the type of photograph which would be obtained with a motion picture camera. Other still cameras, while not contributing to data collection, may be employed to photograph subjects, apparatus and the testing site to furnish detailed illustrations for research records and published reports.

16-MM CINEMATOGRAPHY

As mentioned earlier, the 16mm movie camera is the most frequently used in biomechanics research. It is therefore important to consider in more detail the relation between camera, lens and film specifications and effectiveness in the research application.

16mm Cameras. Although 16mm cameras are available in both spring and motor-driven models, the latter are preferred. Motorized cameras are generally capable of maintaining more consistent frame rates than are the spring-wound variety in which the speed is influenced by the tautness of the spring as well as temperature and humidity variations. The motor may be powered by an alternating or direct current source, depending upon the design of the camera. Direct current models which can be operated from a battery have an advantage if a great deal of filming is to be done at sites where electrical outlets are not available.

The camera should be able to provide a variety of exposure times in order to accommodate sports skills of differing speeds. The faster the action, the shorter must be the exposure time otherwise blurring will make it difficult to distinguish the exact location of a body landmark. Exposure time represents the length of time that light is permitted to expose the film and is, therefore, a function of both frame rate and shutter opening. Most spring-driven cameras have a maximum rate of 64 frames per second and a fixed 180-degree shutter. This means that it takes $\frac{1}{64}$ second for the frame to be moved in front of the aperture, exposed and then advanced so that the succeeding frame can be exposed. If the shutter is pictured in its simplest form (Figure 6-2) as a disc with a 180-degree arc removed

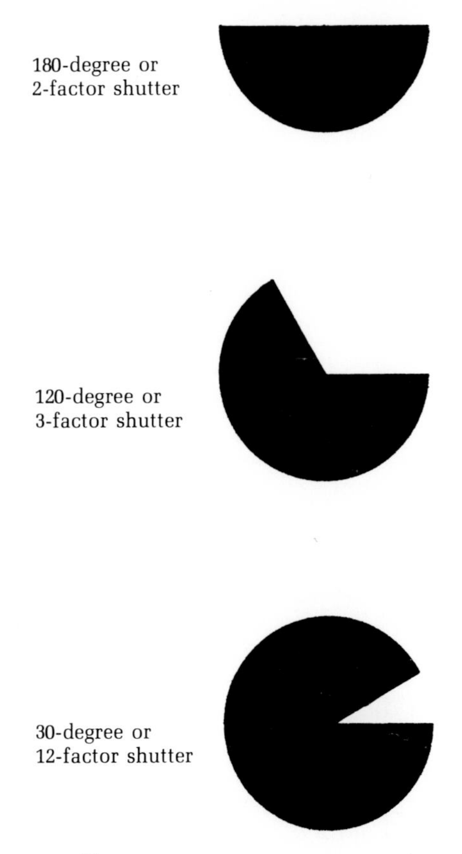

FIGURE 6-2. *Camera Shutters.*

which rotates once while the film is in the gate,† then the resulting exposure time is:

$$\frac{1}{64} * \frac{180}{360} = \frac{1}{128}\text{sec} = .0078 \text{ sec.}$$

Thus, for every .0156 second that the frame is in the gate, light only reaches it for .0078 second or half the time. Part of the remaining .0078 second is utilized for transporting the film into and out of the gate. This conventional frame rate and permanent shutter combination is too restrictive to adequately film such sports skills as the high jump, golf swing, tennis serve and baseball pitch. A shorter exposure time can be obtained with the more sophisticated motorized cameras which have variable or interchangeable shutters and which are capable of much higher frame rates. For example, the exposure time

† During the time that a frame is being exposed, it is held in a film channel referred to as the gate.

may be reduced to $\frac{1}{2400}$ second by using a 30-degree shutter and a camera speed of 200 frames per second. It should be realized, however, that as the exposure time is decreased, a greater amount of light is required for the filming.

Although there are high-speed cameras which can achieve rates in excess of 5,000,000 frames per second, applications in biomechanics seldom require more than 500. Intermittent cameras, in which the film remains stationary in the gate during the exposure, may be capable of 400 to 500 frames per second. In this speed range, registration pins are necessary to hold the film in position while the light impinges upon it. Without pin registration, the film would continue to move due to its own inertia and would buckle in the gate area. Although limited in the maximum frame rate which they can achieve, intermittent cameras provide much clearer images than are possible with higher-speed cameras utilizing a rotating prism mechanism which necessitates continuous film movement during the exposure. Table 6-1 presents a summary of the more important factors to consider in choosing a camera for biomechanics research.

Lenses. The complexity and quality of lenses vary extensively. A simple converging lens is assumed to have negligible thickness and is represented diagrammatically by a single point designating its center. Thus, the distances from the lens to the image and from the lens to the object being photographed are measured from this point. Each lens has a principal (or optical) axis about which it is symmetrical. When light rays from a long distance (approximating infinity) enter a simple converging lens, they are deflected inward and intersect at a point on the optical axis known as the principal focus. The focal length of such a lens is then defined as the distance from the center of the lens to the principal focus.

Compound lenses, which are usually corrected to reduce distortion, have a thickness which cannot be disregarded. As a consequence, the lens-object and lens-image distances must be measured from the front or first and rear (second) nodal points, respectively, rather than from the lens center. The effective focal length of the lens is measured between the rear principal focus and second nodal point or between the front principal focus and the first nodal point. The same results are achieved either way.

Regardless of whether a simple or compound lens is being discussed, the following relationship applies:

$$\frac{1}{u} + \frac{1}{v} = \frac{1}{f}$$

in which u is the lens-object distance, v is the lens-image distance

Table 6-1. General Specifications of 16mm Cameras

Specification	*Comments*
Camera Drive	Spring or motor (AC-DC) driven?
Type	Intermittent film transport or rotating prism?
Pin Registration	Prevents image drifting when film is projected.
Sampling Rate	Minimum and maximum frame rates possible? Fixed or continuously variable?
Speed Regulation	Time to reach designated frame rate? Consistency of frame rates?
Image Quality	Resolution and steadiness of image?
Lens Mount	Types of lenses accepted? Single lens or revolving turret?
Shutter	Fixed, variable or interchangeable?
Focusing and Viewing	Through-the-lens reflex viewing?
Ease of Loading	Film magazines? Self-threading?
Footage Indicator	Self-setting?
Internal Timing or Event Marker Lights	Availability? Number? Location with respect to the gate?
Operational Environment	Effect of temperature and humidity variation?
Film Capacity	100, 200 or 400 feet? Can smaller reels also be used?
Weight and Dimensions	Influence portability.
Construction	Sturdy or easily damaged?
Warranty and Repair	Availability of parts and repair facilities?
Cost	Can optional features be added later?

and f is the focal length or effective focal length. It should be noted that the manufacturer's specification of the focal length, which appears on the lens barrel and applies when objects are focused at infinity, is only accurate to within ±4% of the true value (Focal Encyclopedia of Photography, 1969).

When selecting a lens for a particular filming situation, the researcher ought to give first consideration to its focal length. An understanding of fundamental optical principles will guide him in making the proper selection. In the illustrations which follow, a simple lens is presented and focal length is assumed to be equal to the lens-image distance. The same concepts, however, apply to a compound lens system.

In Figure 6-3, rays of light travel from object O through the center of the lens. The distance between the lens and image I is given by focal length f, while the lens-object distance is specified as D. The triangles to the right and left of the lens are similar. Therefore, the following equation is derived:

$$\frac{I}{f} = \frac{O}{D}.$$

When rearranged, it indicates that: $I = \frac{O * f}{D}$ and $O = \frac{I * D}{f}$.

It can be seen that by increasing the focal length, the size of the image will increase provided the other factors remain constant. If O is taken to represent the width of the field being photographed, it is also evident that by increasing the focal length, the field width is decreased (Figure 6-4). Thus, for a given lens-object distance, a

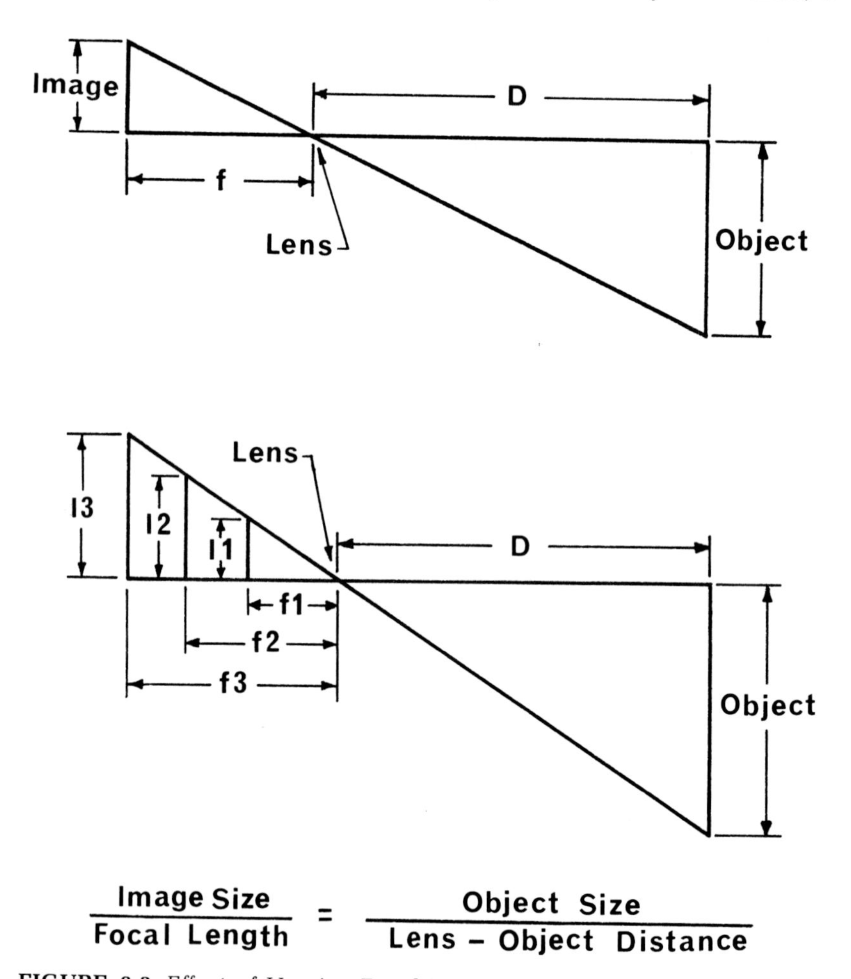

FIGURE 6-3. *Effect of Varying Focal Length on Image Size.*

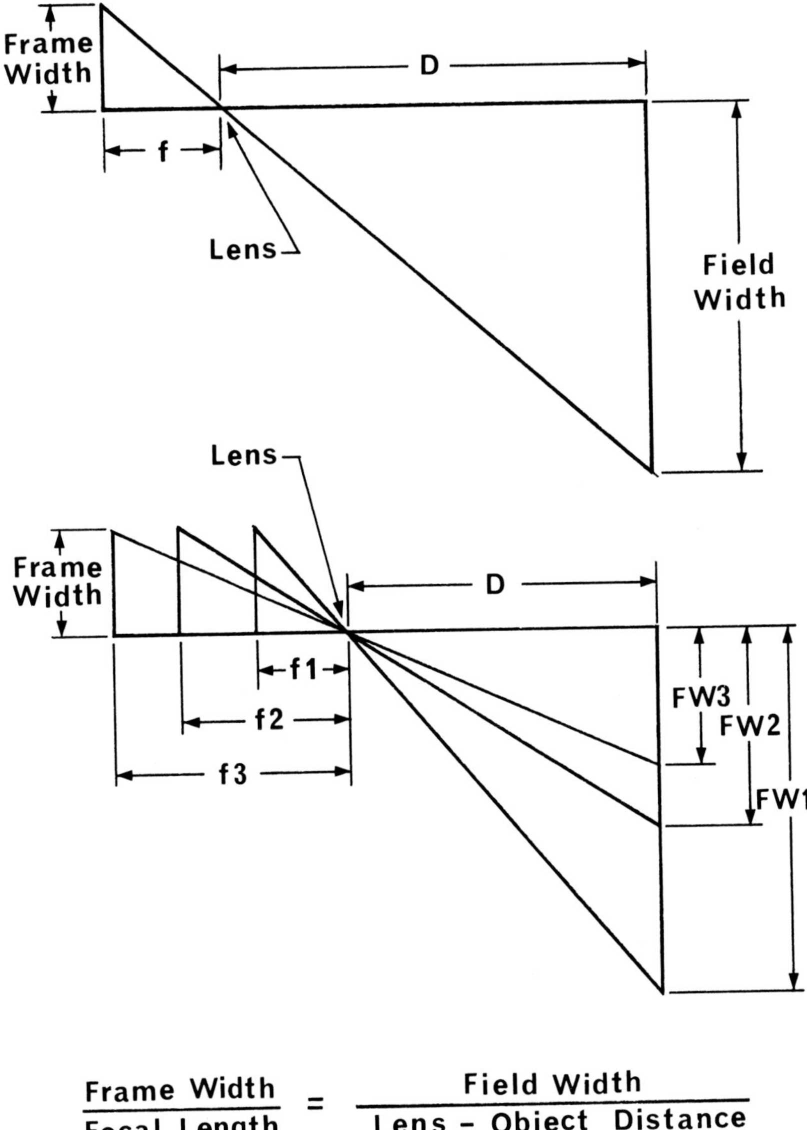

$$\frac{\text{Frame Width}}{\text{Focal Length}} = \frac{\text{Field Width}}{\text{Lens - Object Distance}}$$

FIGURE 6-4. *Effect of Varying Focal Length upon Field Width.*

wide angle or short focal length lens (13mm to 17mm) will photograph a larger field than will a telephoto or long focal length lens (50mm or longer). To cover the same field with a 25mm (normal focal length) and a 50mm lens, the camera with the 50mm lens would have to be positioned twice as far back from the object.

When quantitative data are to be taken from film, it is important that the image be as large as possible with respect to the frame. This can best be achieved by using a long focal length lens. Prior to the actual filming, the image size can be estimated from a knowledge of the focal length, lens-object distance, object size and the magnification provided by the projector to be used in the analysis.

It is also desirable to minimize the perspective error which occurs when parts of the body or sports implements lie outside the principal photographic plane.† As a result, the image of the arm or leg closer to the camera will be larger than that of the corresponding limb on the opposite side of the body even though they may be identical in real life (Figure 6-5). This type of error is also present when a limb is at an angle to the photographic plane and hence appears shorter than when it is situated exactly within the plane. Although perspective error cannot be completely eliminated, it can be reduced by positioning the camera as far back from the performer as possible (Table 6-2; Figure 6-6). The increased lens-object distance will not be accompanied by any loss in image size provided a long focal length lens is used. It should be noted that, contrary to popular belief, movements which occur anywhere in the selected photographic plane perpendicular to the optical axis of the camera will not be subject to perspective error (Figure 6-7).

Parallax error, which refers to the apparent discrepancy in the location of two objects, occurs when the objects are observed from different points. Such an error is experienced when the camera is not equipped with through-the-lens reflex viewing. As a result, the positions of close objects seen by the lens and an off-set viewfinder are not identical. At lens-object distances exceeding six or seven feet, however, this discrepancy is not serious (Focal Encyclopedia of Photography, 1969). In cinematography, therefore, parallax need not concern the researcher since he is dealing with relatively large lens-object distances. A possible exception might be the filming of a timer or chronoscope at close range. In this case, care would have to be taken to be sure that the object was actually included in the photographic field.

In most 16mm cameras, lenses can be interchanged to suit the requirements of a particular experimental situation. The 25mm is probably the most commonly used in biomechanics research. It will provide about a 16-foot field when the camera is located 40 feet from the subject. While longer focal length lenses are desirable, their use generally requires a lens-object distance in excess of 60 feet. Although this probably could be accommodated at an outdoor filming

†This is the major plane in which the subject is performing. It is always perpendicular to the optical axis of the lens.

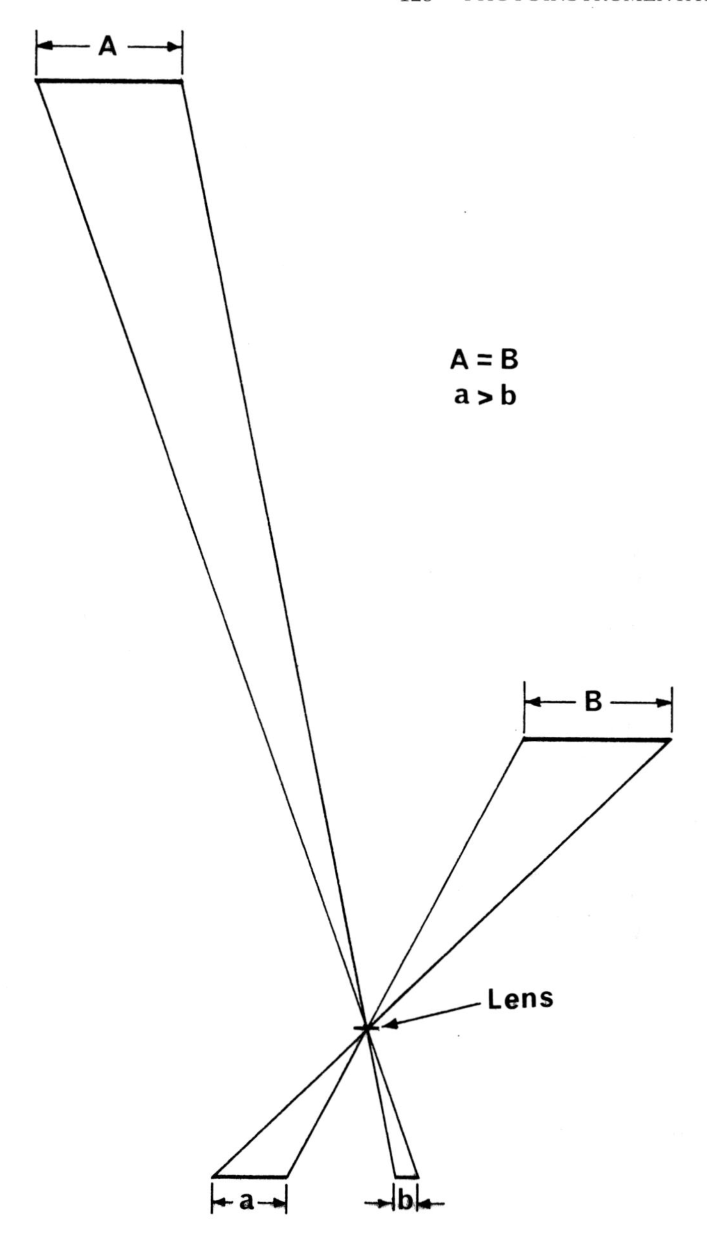

FIGURE 6-5. *Inverse Relationship Between Image Size and Lens-object Distance.*

site, it might not be feasible in a gymnasium or swimming pool. A zoom or variable focal length lens may provide a solution to this dilemma. In the past, zoom lenses have been considered inferior because of the presence of distortions at certain positions within the

Table 6-2. The Inverse Relationship between Perspective Error and Lens-object Distance

$$I = \frac{O * f}{D}$$

Object Size O (*ft*)	*Focal Length* f (*mm*)	*Camera-object Distance* D (*ft*)	*Film Image* I (*mm*)
10.00	15.00	15.00	10.00
10.00	15.00	16.00	9.38
10.00	45.00	45.00	10.00
10.00	45.00	46.00	9.78
10.00	90.00	90.00	10.00
10.00	90.00	91.00	9.89

range of the zoom. With the better quality lenses now being manufactured, however, this does not present as serious a problem as it did previously. In all cases, lenses of high quality should be employed so that lens aberrations can be reduced to a minimum.

In addition to focal length, lens specifications indicate the light transmission capability. This is referred to as the f/ number† or relative aperture and is the ratio of the focal length to the aperture diameter. For example, an f/1.9 lens is wide open and will transmit the maximum amount of light to the film when its aperture diameter is 1/1.9 of the focal length. The diaphragm of the lens can be controlled to regulate the amount of light reaching the film. Readings from an exposure meter which account for film sensitivity and exposure time denote the appropriate f/ number to be set on the lens. Under low ambient light conditions as might be encountered in a swimming pool, the lens opening should be fairly large (f/1.9 or f/2.2). In contrast, if the filming takes place in bright sunlight, a much smaller aperture would be required. An f/22 where the aperture diameter is 1/22 of the focal length might be indicated by the exposure meter. It should be noted that when a lens is stopped down so that there is only a small opening, the depth of field in sharp focus will be increased.

Before filming, the lens-object distance must also be set to ensure a well-defined image. When the object is close to the camera, this setting is critical. Lenses will vary in terms of the minimum distance

†The slash in f/ denotes a fraction. Thus, higher f/ numbers such as 16 or 22 indicate smaller diaphragm openings than do numbers such as 1.9 or 2.8.

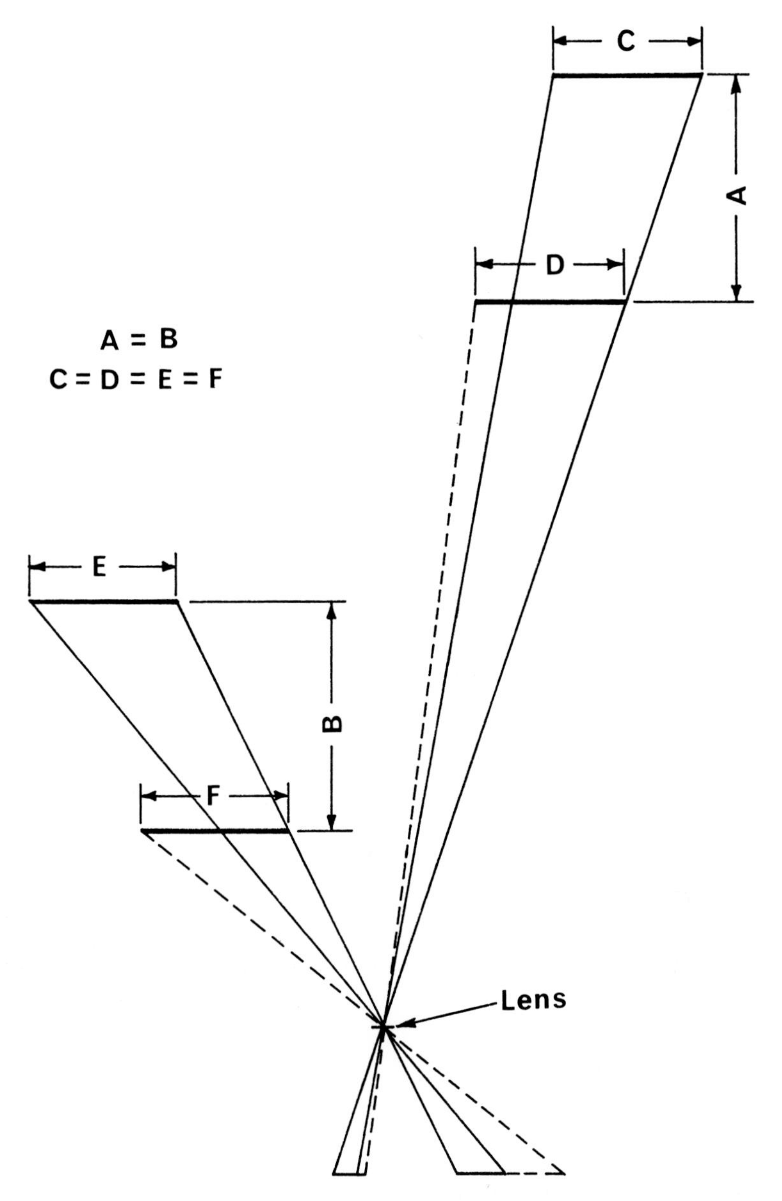

FIGURE 6-6. *Reduction in Perspective Error with Increased Lens-object Distance.*

at which an object can be focused. In most cinematographic studies in which the camera is positioned well back from the photographic plane, a reasonable estimate of the lens-object distance will usually result in acceptable pictures.

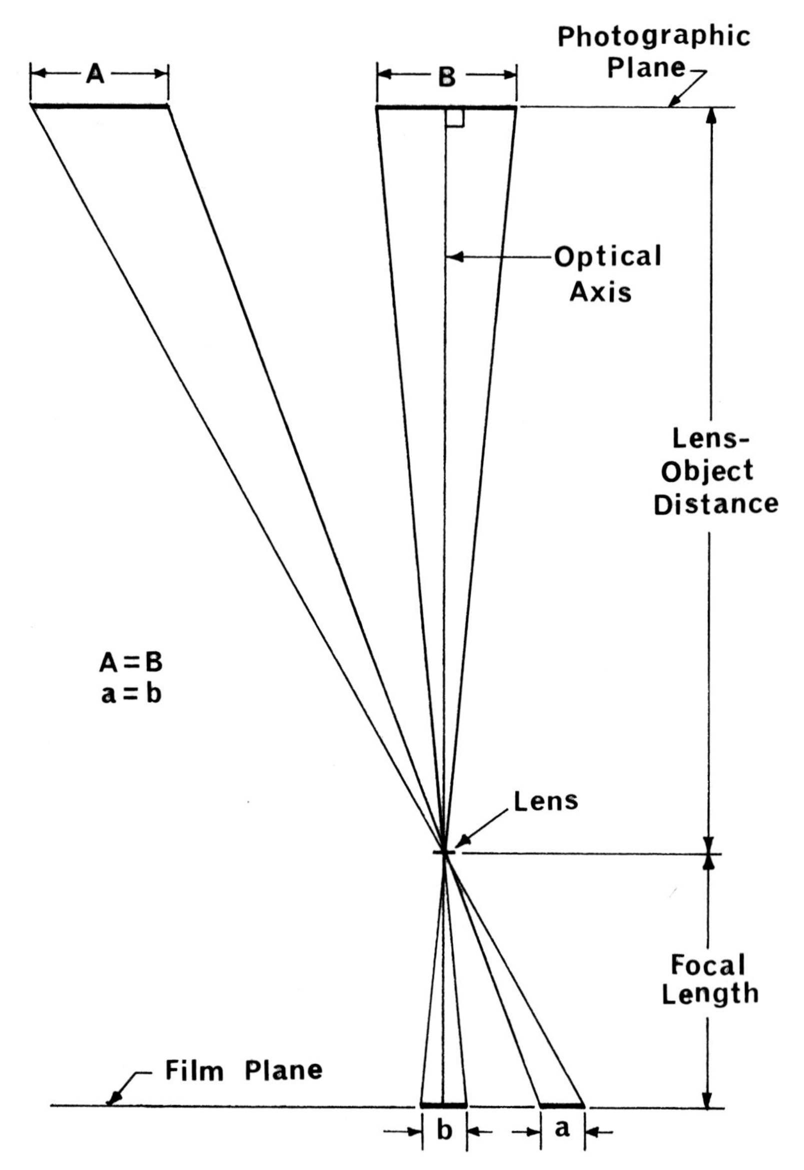

FIGURE 6-7. *The Generation of Equal Size Film Images by Equal Lengths Anywhere in a Plane at Right Angles to the Optical Axis.*

Film. Films are rated according to their light sensitivity by the American Standards Association. Each type has an ASA Index, with the higher numbers indicating faster films which are more sensitive to light. When selecting film for research purposes, a compromise must be made to achieve the best combination of speed, grain size and resolution. Black and white film is generally preferred to color because it is cheaper, requires less light to expose, is easier to process and provides more clearly defined images. Reversal film, in which the original exposure and final positive image are recorded on the same emulsion, has the advantage of being less expensive than negative film and furnishes a more sharply defined image for precision motion analysis. Nevertheless, if the researcher wishes to have several copies made, he should use negative film.

In biomechanics, Tri-X Reversal with ASA 200 in daylight and 160 under tungsten lighting and 4-X Reversal with ASA ratings of 400 and 320 respectively are the most commonly used. These are both triacetate base films. Although faster types which permit pictures to be taken under more restricted light conditions are available, their coarse grain detracts from the image and makes quantitative analysis more difficult. In some instances, it may be advisable to use a polyester (Estar) base film because of its greater dimensional stability. It is also less brittle and hence especially good in cold environments. Another advantage lies in the fact that it is thinner, making it possible to get approximately 25% more footage on the reel. The pressure plate within the camera may have to be modified, however, to prevent movement of this thinner film in the gate.

When ordering film, it is wise to specify not only the type desired but also that it be on daylight loading reels and have perforations on both sides. Double perforated film which must be used in most high-speed cameras also provides greater flexibility in projecting the film for analysis. The number of feet required for a study may be estimated as follows:

$$\frac{\text{Number of Trials} * \text{Time per Trial} * \text{Frame Rate}}{40}$$

with the divisor indicating that there are 40 frames in each foot of 16mm film. Allowance should also be made for a leader and a trailer.

Processing should not present much of a problem in the larger urban centers which have commercial photographic laboratories. Local television stations may also be utilized as they often provide very rapid service in developing 16mm black and white film. Unless a biomechanics researcher is doing an extremely large amount of filming on a continuous basis, it is usually not economically feasible for him to have his own film-processing equipment.

Exposure Meters. As indicated previously, an exposure meter must be employed to measure the light intensity and indicate the proper lens opening (f/ stop) for a given ASA and exposure time. Both incident and reflected meters are available. In the first case, the light incident or hitting the subject is recorded. Thus, the light-sensitive opening on the meter is pointed away from the subject and toward the camera. The reflected type, as the name suggests, indicates the intensity of light reflected from the scene to be photographed. Therefore, it is directed toward the subject when the reading is made. The biomechanics researcher must take care not to be overly influenced by the sky or artificial lighting or spuriously high readings will result. To avoid such errors, the light meter is usually tilted downward at a slight angle. In addition to the two principal types of meters mentioned, some are convertible and can be used for either incident or reflected light. Since the precise details of operation vary from one model to another, the instruction sheet received upon purchase should be retained and followed carefully.

Light meters will use either cadmium sulphide cells which regulate current from small batteries in proportion to the intensity of light being received or selenium cells which convert light energy into electric current. The former are more sensitive, but since they are battery-operated, must always be checked to see if the battery is at full strength. A device should be included on such meters for this purpose. Although the selenium cell light meters are not quite as sensitive, they should provide true readings for the life of the instrument (Time, 1970).

CALIBRATION OF CAMERA SPEED

Since time data play an integral role in the calculation of velocity and acceleration, every effort should be made to ensure their accuracy. Time is derived from the frame rate at which the camera is operating. Thus, if the camera speed is 100 frames per second, the time per frame is .01 second. Although 16mm cameras have a dial or indicator on which the desired frame rate is set, this value is not considered sufficiently accurate for quantitative film analysis. The speed of spring-wound cameras may be influenced by conditions of temperature and humidity as well as by the tension in the spring. Motor-driven cameras tend to operate at a constant speed but there may be a deviation of ± 10 frames from the one specified on the indicator. Therefore, adequate calibration is essential.

One of the earliest methods of estimating camera speed was to film a freely falling object which was unaffected by air resistance (Cureton, 1939). The object was released at a known distance above the ground with the elapsed time during the drop being computed from the equation of motion:

$$s = v_i t + \tfrac{1}{2} a t^2$$

Thus, v_i (initial velocity in the vertical direction) = 0
s (vertical displacement) = a known quantity in feet
a (constant acceleration due to gravity) = -32.2 ft/sec/sec.
The number of frames over which the action occurred, combined with the calculated time, permitted the determination of the frame rate. This technique had several limitations, and at best, gave only a rough approximation of the actual camera speed. The greatest problem was in identifying the exact frame in which the object was released and the one in which it completed the specified distance.

Although a chronoscope or a stopwatch may be photographed before or after the action, it is desirable to have an indication of the frame rate while filming the actual performance. A rotating cone timer (Blievernicht, 1967) and a dichroic mirror arrangement (Cooper and Sorani, 1965) have been suggested for this purpose. Large sweep-hand clocks have also been included in the field of view. The most accurate method to date, however, appears to be the digital timing display developed by Walton (1970). Four rows of 10 neon lights are employed, with a row to indicate the full second, the $\frac{1}{10}$ of a second, the $\frac{1}{100}$ of a second and the $\frac{1}{1000}$ of a second respectively. Since only one bulb is on in a row at any instant, the time may be read directly. Two or more of these timing display units can be driven simultaneously by a stable quartz crystal oscillator. Subsequent modification by Petak† resulted in a binary coded decimal unit with four bulbs per row. This reduced the expense of constructing the timer. Unless the camera is operating at high frame rates, however, the thousandth of a second row cannot be interpreted accurately. This limitation has been eliminated by substituting decimal rather than binary displays for the final two rows of lights (Figure 6-8).

Some of the more sophisticated cameras are equipped with small internal lights which expose the side of the film in response to an

†K. L. Petak, Research Engineer, Biomechanics Laboratory, The Pennsylvania State University.

FIGURE 6-8. *Combination Binary and Decimal Timing Display.*

input signal. When connected to a timing impulse generator, the lights can be pulsed to mark the film a set number of times (1, 10, 100, 1000) per second. These lights can also be employed as event markers to synchronize an external occurrence of interest with the film record. It should be realized, however, that the lights are not located in the gate area. The exposed frame and the timing or event mark are out of phase by a set film distance, depending upon the location of the lights with respect to the gate and the size of the loops left during the threading. The precise distance can be determined by counting the number of frames between the overexposed frame in the gate when the camera first begins to operate for any filming sequence and the corresponding cluster of marks resulting from continued exposure by the lights on one area of the film.

FILMING FUNDAMENTALS

When photoinstrumentation is used in biomechanics research, certain procedures must be carefully observed if accurate data are to be obtained. Unless the films are taken correctly, meaningful quantitative analysis is difficult if not impossible. The following factors should be considered when planning and implementing a cinematographic study.

1. The performer should move at right angles to the optical axis of the camera. In the case of the long jump, for example, the camera must be sighted perpendicular to the runway and pit. Since the projected image on the film is in two dimensions, any part of the movement occurring out of this plane will not be accurately represented as it will be subject to perspective error.

2. The camera should be positioned as far away from the action as possible to minimize the effect of perspective error and a telephoto lens should be used to increase the size of the image.

3. For most research applications, a stationary camera position is required. The camera should be sighted on the center of the action, leveled and secured to a stable tripod. Panning is not recommended because it complicates the subsequent film analysis.

4. The background should be plain, uncluttered and provide a contrast to the subject. Shiny surfaces should be avoided as they tend to cause undesired reflections. To orient the film consistently during the analysis, horizontal and vertical references should be included if none exist in the natural background. The photographic field should also include such identification information as subject number, trial, experimental condition, date and test session. Swimming or track lap counters or gymnastic score indicators can be used for this purpose (Figure 6-9).

FIGURE 6-9. *Subject Markings and Photographic Background (Barlow, D. A.: Relationship between Power and Selected Variables in the Vertical Jump. Unpublished Master's Thesis, Pennsylvania State University, p. 79, 1970).*

5. The subject should be familiar with the conditions of the study and understand what is required of him. A concerted effort should be made to provide a "normal" setting for the filming.

6. To facilitate the location of segmental endpoints during the analysis of the film, the subject should avoid bulky clothing which obscures the joint markings. Where possible a swimsuit or leotard should be worn. In activities like skiing and skating, the subject should wear tight clothing which does not move relative to his skin during the performance. Contrasting markings should be placed on the joint centers or around the joints. Felt pen, tape or self-adhesive dots can be used. If the instant of foot contact is to be detected, there must be a definite contrast between the shoe and the floor or track. Sports implements used in the performance should also be marked appropriately.

7. There must be sufficient light to expose the film. When shooting indoors at high frame rates, artificial light is often required. It should illuminate the subject evenly but not distract him. Under competition conditions during which artificial lighting is prohibited, adequate exposure may be obtained by using faster film, reducing the frame rate, increasing the shutter opening or using some combination of these three.

8. The frame rate selected should be sufficiently high to eliminate blurring of the most rapidly moving part of the body.

9. A spring-driven camera must be rewound tightly before each trial. This will help to ensure a more consistent camera speed for each performance.

10. The camera should be started a few seconds prior to the beginning of the action to allow sufficient time for it to reach the desired frame rate.

11. Provision must be made to calibrate the speed of the camera either by including an accurate timing device in the photographic field or by utilizing internal timing lights.

12. It is essential to establish a scale value which will permit the conversion of film measurements to real distances. This can be done by photographing an object of known length in the plane of the motion either prior to or following the performance.

13. Conditions, distances and other important details of the filming should be recorded on a form similar to the one shown in Figure 6-10. This information will prove invaluable when documenting or replicating the study.

QUANTITATIVE FILM ANALYSIS

Although film has been used to investigate various patterns of human motion for almost a century, advances in film analysis pro-

FILMING RECORD

Purpose ______________________ Date ______________________

Investigator ______________________ Location ______________________

Time Started ______________________ Time Completed ______________________

Camera ______________________ Frame Rate __________ Shutter __________

Exposure Time ______________ Lens ______________ f/ Stop ______________

Light Conditions ____

Artificial Lights (number & type) ______________________________

Background ____

Reference Marks ____

Identification ____

Camera Speed Calibration ______________________________________

Camera Height ______________________ Camera-Subject Distance __________

Diagram of Filming Site

Subjects (number, joint markings, etc.) ______________________________

____

____

Filming Sequence (Scale, trial order, etc.) ____________________________

____

____

Comments ____

____

____

FIGURE 6-10. *Sample Filming Record Sheet.*

cedures have tended to lag behind other areas of biomechanics research. This problem has now been at least partially rectified by the development of modern motion analyzers and the application of computer technology in processing the film data.

Motion Analyzers. One of the earliest techniques used to extract position data (Marey, 1902), which is still employed to some extent, requires only a stop-action projector and a flat tracing area. The film image is projected directly onto a wall or off a 45-degree mirror onto a horizontal translucent tracing surface. Important body landmarks and segmental endpoints from successive frames are recorded on graph paper attached to the tracing surface. Lines connecting these points reconstruct a composite outline of the motion from which linear and angular measurements can be made. When using this technique, the biomechanics researcher must take care that the projection is at right angles to the drawing surface, otherwise the image will be distorted. It is also necessary to mark stationary background references on the paper to ensure consistency in positioning the frames. Although this method requires a considerable length of time to obtain quantitative information from the film, it does provide a large image and can be implemented with a minimum of expense provided a stop-action projector with single-frame advance is available.

A second technique utilizes a Recordak Film Analyzer in which the magnified image is projected onto a ground glass screen. This unit does not have a frame counter and the film must be turned between the supply and take-up reels by hand. Thus, for composite tracings, it is particularly important to be certain that the frame orientation does not change when the film is advanced. Because of the difficulty in keeping track of the elapsed frames essential for establishing a time base, the laborious task of recording the data and the inherent inaccuracy in making quantitative measurements from composite tracings, the Recordak is not recommended for precision motion analysis. It is suitable, however, for student projects and laboratory exercises of an elementary nature. Recordak units can often be located in the microfilm sections of university libraries.

More sophisticated motion analyzers (Table 6-3) are now available and are being widely used to study the mechanics of sports skills. One such analyzer manufactured by Vanguard Instrument Corporation consists of a projection head and a projection case (Figure 6-11). The projection head rotates freely about a vertical axis, making it possible to align the frame horizontally and vertically. A notable feature of the Vanguard Analyzer which is not found on the typical stop-action projector is pin-registration. As the film advances, it is locked into place to assure the same relative position from frame to frame. Provision is also made for minor horizontal and vertical adjustments of image position to accommodate image drift originating in non-pin-registered cameras. The motor-driven projector will transport the film forward or backward either continuously or frame by frame. As each frame passes through the gate, a counter

Table 6-3. General Specifications of Motion Analyzers

Specification	*Comments*
Image Magnification	A large image is desirable for analysis. Are lenses of different magnification interchangeable?
Film Transport	Motorized? Frame by frame or free flow? Forward? Reverse? Single frame? Multiple Frame? Ciné?
Film Capacity	Size of reel accomodated? 400 feet?
Position of Projected Image	Is the image projected from the rear, above or below?
Level of Illumination	Is a darkened room necessary for analysis?
x and y Coordinates	Are these incorporated into the system? What is their accuracy? Is the origin of the coordinate system fixed or can it be changed?
Coordinate Readout	Dials? Computer cards? Computer tape?
Provision for Angle Measurement	Can angles be read directly from the analyzer?
Adjustment of Image Position	Will projection head rotate? Can the image be shifted horizontally and vertically for minor adjustments?
Pin Registration	Is the frame secured in place by register pins to prevent movement of the image during analysis and provide consistent orientation?
Frame Counter	Is one included? Can it be reset to zero? Is it visible to the operator during analysis?
Computer Compatibility	Can the coordinate readout be automated later if desired?
Availability of Parts and Repair Service	
Cost	

is tripped so that a specific number can be associated with any given frame. Lenses which magnify the image between 16 and 27 times its original size can be interchanged to meet the needs of a particular study. Either a 16mm- or a 35mm-projection head may be used, depending upon the dimensions of the film being analyzed.

The projection case provides a rear-projected image upon which a pair of movable crosshairs are superimposed. Two control wheels at the bottom of the case permit the operator to manually position

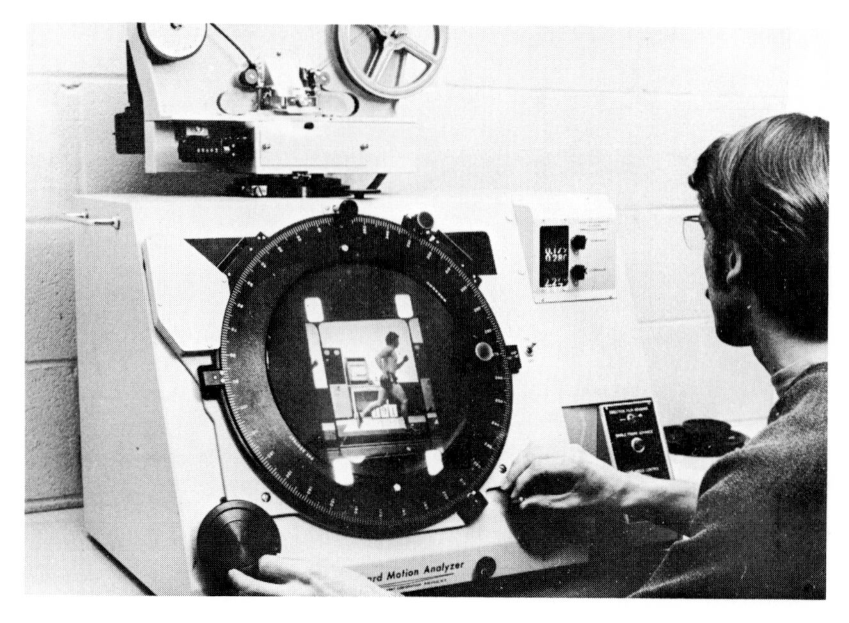

FIGURE 6-11. *Vanguard Film Analyzer.*

the crosshairs so that they intersect at the center of some desired point on the film image. The x and y coordinates of this point are then read from the dials at the side of the machine. These values represent distance measured to the nearest $\frac{1}{1000}$ of an inch on the projection screen. In addition to the standard features, an angle screen can be purchased to permit direct measurement of angular position.

Within the past decade, the high-speed digital computer has been accepted as an essential partner in the processing of large volumes of data obtained from the motion analyzer. The coordinate and time values from the film are customarily recorded on data sheets and subsequently punched on computer cards or tape. Elimination of the intermediate transcription stage is possible by having a key punch or teletype adjacent to the analyzer. With this arrangement, the operator can position the crosshairs over the designated points on the film image and a co-worker can put them directly on the cards or tape used as input to the computer program (Figure 6-12). Even more labor can be saved by having the analyzer itself automated so that the mere pressing of a button will result in recording specified information on the computer input medium (Figure 6-13).

In some cases, the conventional motion analyzer is being replaced by a computerized graphic tablet or platen onto which the film image is projected. One such system (Figure 6-14), called the Graf/Pen, is based upon the principle of sound transmission. The operator uses

FIGURE 6-12. *Semi-automated Film Analysis System.*

FIGURE 6-13. *Automated Film Analysis System.*

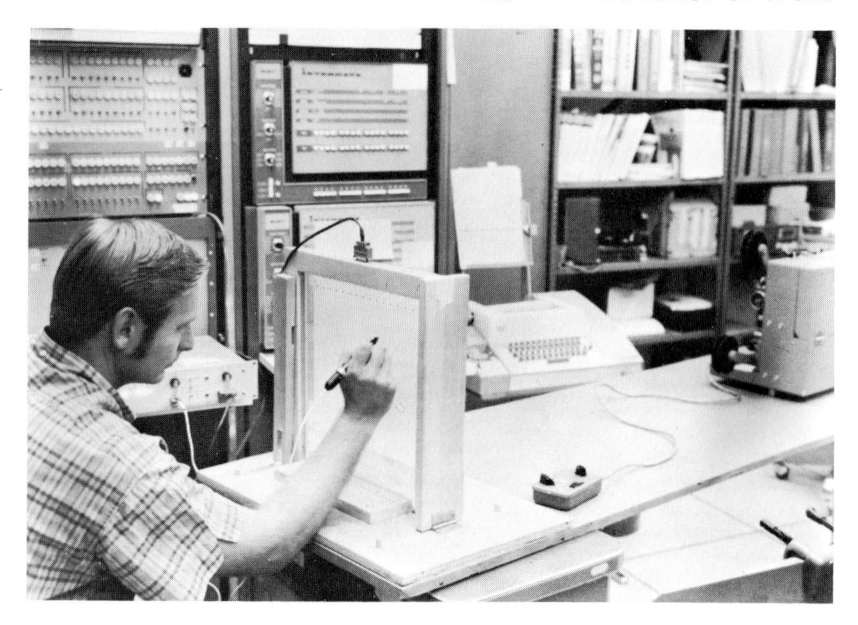

FIGURE 6-14. *Automated Data Tablet* (*Courtesy of Marlene Adrian, Ph.D., Washington State University, Pullman, Washington*).

a special metal stylus which creates a spark when it contacts the tablet. The sound generated by this contact is picked up by two strip microphones located at right angles to one another on the edge of the tablet. The x and y coordinates of the spot touched by the stylus are then determined as a function of the time required for the sound to reach the microphones. With the addition of a suitable interface, this system can be coupled directly to a computer, magnetic tape recorder or teletype.

Almost complete automation for film analysis has been achieved by using a flying spot scanner (Groh and Baumann, 1968; Kasvand, 1970; Kasvand *et al.*, 1971). With this instrument, an electron beam sweeps back and forth across the film image to detect specified levels of light intensity. It can thus distinguish dots placed on the segmental endpoints provided they contrast sufficiently with the subject and the background. The scanner is interfaced to a computer which is programmed to calculate the coordinates of the designated points. Some difficulties still exist, however, in differentiating one landmark from another if they overlap, as would occur when the hand passes the hip during a walking cycle (Quanbury, 1972).

Fundamentals of Film Analysis. Since the position and time information obtained from the film provide the raw data for the calculation of displacements, velocities and accelerations, it is important to strive for a high degree of accuracy in determining their values. The observation of certain fundamental film analysis procedures will facilitate the achievement of this objective and will result in a more efficient approach to the analysis.

1. A preliminary examination of the films should be made to determine the specific trials and the number of frames in each trial to be analyzed. Unless the camera speed was extremely slow in relation to the action, it is not necessary to analyze each frame. Every third or fourth frame may be sufficient, depending upon the amount of movement which occurs during the interval. Frames of special significance such as the take-off and touch-down in running should also be analyzed.

2. An efficient method for obtaining the necessary information from the film should be developed first by using a single trial. The order in which the coordinates are recorded should be one involving a minimum of extraneous movement on the part of the operator. Efforts should be made to record data from which several different variables can be calculated. For example, if the data are to be processed by computer, a program can be written to determine the mass center, linear and angular limb positions and joint angles from coordinates of the segmental endpoints. If the calculations must be done by hand, angular measurements should be made directly from the film image.

3. The first frame of each motion sequence which is analyzed should be identified for future reference. This can be easily accomplished by marking the corner of the frame with a fine felt tip pen. If a timing display is included in the photographic field, the time of the first frame should also be recorded.

4. The investigator must become proficient in locating the segmental endpoints or joint centers of the body on the film image. Carefully placed markings on the subject prior to the filming will make this task easier.

5. When making angular measurements with a protractor or reading them directly from an angle screen, it is not necessary to compensate for the reduced size of the image. The angles, however, must be in the plane of the projection; otherwise, their true values will not be obtained. A computer program could also be written to calculate angles from the coordinates of the segmental endpoints. Such a program would include a system for identifying the relative magnitude of the angle (0-90°, 90-180°, 180-270° or 270-360°) and would apply the cosine law (Figure 6-15). The measurement of body angles over successive frames in conjunction with a knowledge of the elapsed time makes it possible to determine angular velocities and angular accelerations. The angle of projection of an object can

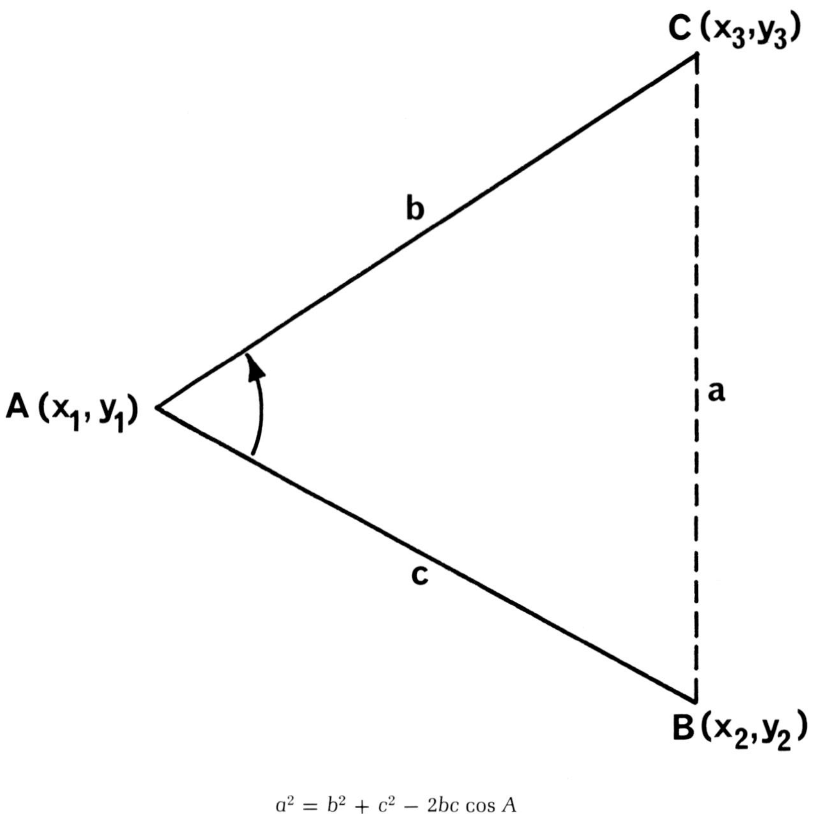

$$a^2 = b^2 + c^2 - 2bc \cos A$$

$$A = \text{arc cos}\left(\frac{b^2 + c^2 - a^2}{2bc}\right)$$

in which

$$a = \sqrt{(x_3 - x_2)^2 + (y_3 - y_2)^2}$$
$$b = \sqrt{(x_3 - x_1)^2 + (y_3 - y_1)^2}$$
$$c = \sqrt{(x_2 - x_1)^2 + (y_2 - y_1)^2}$$

FIGURE 6-15. *Calculation of an Angle from Film Coordinates.*

be approximated from the coordinates of its mass center in the frames immediately preceding, at and following release.

6. Linear displacements from the film must be converted to life size since the projected dimensions will be considerably smaller than their true size. This adjustment can be made by multiplying film distances by an appropriate scale factor obtained from the measurement of an object of known length photographed in the plane of the motion. For example, if a three-foot rod measures .5 units when projected, then

$$.5 \text{ units} = 3 \text{ feet}$$

$$1 \text{ unit} = 6 \text{ feet,}$$

and thus every unit on the projected image would have to be multiplied by a factor of six to convert it to actual life size in feet.

7. The camera speed must be known in order to determine the time per frame used in velocity and acceleration calculations. Any of the previously described methods for calculating frame rate could be used for this purpose. If the camera is operating at 100 frames per second, then the time per frame will be $\frac{1}{100}$ or .01 second; and if it is operating at 50 frames per second, the time per frame will be .02 second. In the latter instance, the time elapsed between frame 24 and frame 36 would be:

$$(36 - 24) * (.02) = .24 \text{ second.}$$

Time per frame should not be confused with exposure time.

8. It is recommended that some type of smoothing be applied to the displacement data prior to calculating velocity and acceleration. As with other experimental data, the differentiation of experimental values tends to produce erratic results. This is particularly evident in the accelerations and in the case where the time intervals used in computing the velocities and accelerations are extremely small. Various smoothing techniques are presented in Appendix C.

THREE DIMENSIONAL CINEMATOGRAPHY†

A major difficulty in studying movement cinematographically has been that of obtaining spatial coordinates. While a single camera may be employed to study a pseudo two-dimensional activity such as running, it will not provide sufficiently accurate data for a complete quantitative description of throwing, high jumping or any other skill incorporating a significant twisting motion. The latter require a technique for determining x, y and z coordinates from film. These coordinates, in turn, serve as the basis for calculating displacement, velocity and acceleration in spatial terms.

Single Camera Techniques. In seeking to avoid the three-dimensional problem, many investigators have chosen to study movements occurring primarily in one plane and have felt justified in considering any movement out of that plane as inconsequential. By using telephoto lenses which permit large camera-to-subject distances, they have attempted to minimize the perspective error in-

†Portions of this section have appeared in *Kinesiology 1973.*

herent in planar analysis. Others have utilized two or three cameras and have conducted an independent analysis of each view.

At the University of Wisconsin, "body belts" were constructed from heavy elastic upon which styrofoam projections or fins were fixed by means of light aluminum holders (Lamaster and Mortimer, 1964). These belts, attached to the pelvic girdle and thorax of the subject, were utilized to investigate spinal rotation during the performance of different sports skills. The angles of rotation were obtained directly from an overhead camera or calculated from the film of a side camera by comparing the length of the styrofoam fin when it was in the picture plane with its apparent length measured from the film. The degree of accuracy of these angular measurements is open to question, however, as the fins tend to vibrate independently of any movement on the part of the subject (Atwater, 1972). Their presence may also have some influence upon the "normal" patterns of motion of the performer.

The general method for estimating the nonplanar angle of the limb by comparing the apparent and true lengths (Figure 6-16) has several limitations. Since the precise distance between the camera and the plane in which the subject is moving may not remain constant, it is difficult to specify the exact scale factor necessary to convert film dimensions to actual size. Even if this can be accomplished satisfactorily and the depth coordinate of one end of the limb estimated with respect to the other, it is almost impossible to determine the absolute depth coordinates of all body points which will be consistent from frame to frame. In addition, the technique is based upon the assumption of parallel projection, which is only tenable when the distance between the subject and camera is extremely large. In spite of Plagenhoef's (1968) proposed correction to overcome the latter difficulty, methods for obtaining spatial coordinates from the film of a single camera have not proven adequate for detailed studies of nonplanar movements.

Mirror Techniques. The inclusion of a mirror within the photographic field may provide a second image of the performer from a different angle. If the optical axis of the lens is positioned at a 45-degree angle to the plane of the mirror, the resulting image will be equivalent to the superimposing of the images from two identical, perfectly synchronized cameras whose optical axes intersect at right angles. While several investigators have utilized mirrors in cinematographic studies, few have calculated x, y and z coordinates from the film. Analysis of the mirror image has been generally limited to the temporal or qualitative levels.

Bernstein (1930), however, developed a method for obtaining

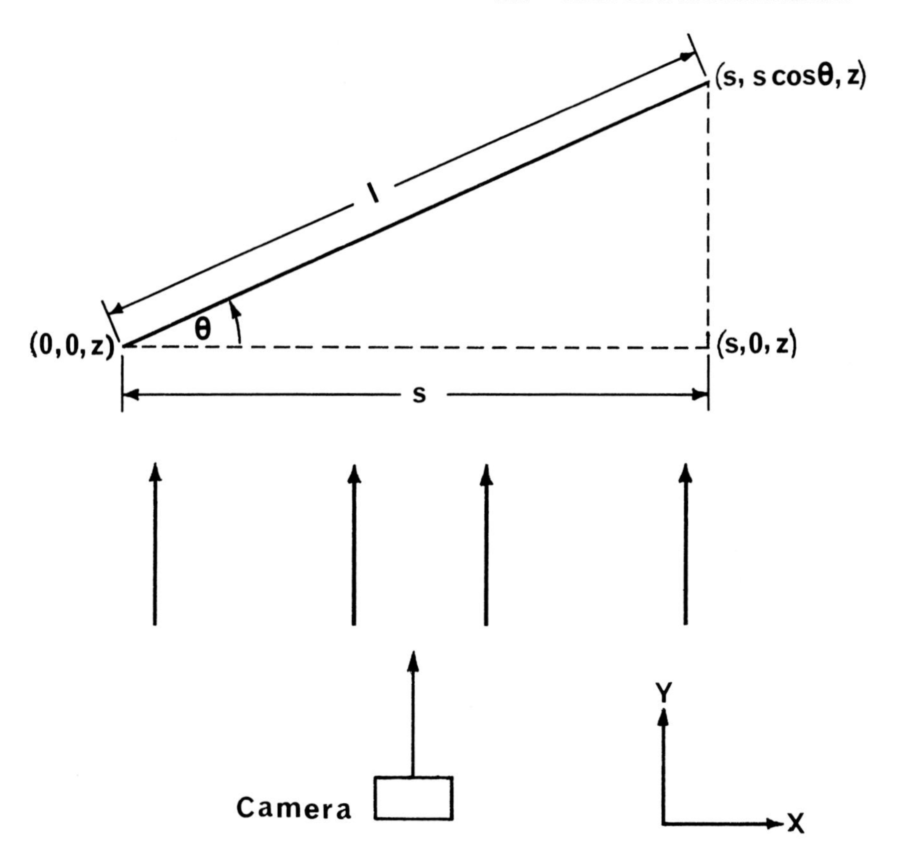

l True length of a segment
s Apparent length of a segment
θ Angle of the segment with respect to the photographic plane

$$\theta = \text{Arc}\cos\frac{s}{l}$$

FIGURE 6-16. *Single Camera Estimation of Nonplanar Angles Assuming Parallel Projection.*

spatial coordinates utilizing one camera and a plane mirror. The origin of his rectangular reference frame was located at the point of intersection of the optical axes of the real and mirror cameras. By taking advantage of similar triangles and tetrahedrons formed by various projections, he was able to develop formulae for determining the x, y and z coordinates of points which were visible both in the mirror and real images.

Although having both images on one frame obviates the problem of camera synchronization, the presence of a mirror usually restricts the range of the subject's movements. In addition, the large mirrors commonly found in dance studios and exercise rooms are rear-silvered. As a result the thickness of plate glass in front of the reflecting surface will undoubtedly introduce some distortion. Front-silvered mirrors, which would provide more accurate data, tend to be quite expensive and rather fragile. While the use of small portable mirrors to reflect an external event related to the performance and large fixed mirrors to provide a second view of the subject is to be encouraged, it should be appreciated that it is difficult, and in many cases, impossible to determine precise spatial coordinates from the resulting films.

Stereometric Systems. The stereometric method of creating the three-dimensional effect in photography employs either a special stereocamera (Cimerman and Tomasegovic, 1970) or two cameras placed side by side with their optical axes parallel to one another. Since an established base distance separates the lenses, each has a slightly different view of the object. Standard equations for the calculation of x, y and z coordinates of desired points on this stereopair (Hallert, 1960; Moffitt, 1967) have been applied to the study of human motion by Gutewort (1968, 1971) and Ayoub and associates (1970). With this method, the base distance between the two lenses is an important factor in influencing the magnitude of the error in estimating the depth coordinate. Although a large base is desirable, there is a limit to which it can be increased while an area in the field of view of both lenses is still maintained. Only within this common area can spatial coordinates be determined for points on the image.

Multiple Camera Methods. A number of researchers have concluded that more than one camera must be utilized if spatial coordinates are to be derived from film. Walton (1970) proposed that four cameras be employed to obtain the spatial coordinates of body parts. He acknowledged that two cameras, positioned with their optical axes at right angles, would furnish the necessary information for determining the coordinates provided that particular points were visible in both of these cameras simultaneously. However, the use of only two cameras would limit the number of body references for which x, y and z coordinates could be specified. The provision of four cameras would make it possible to record most body landmarks in at least one set of two appropriately positioned cameras.

Both Atwater (1970) and Anderson (1970) presented methods for establishing spatial coordinates using three cameras, with one sighted along each of the three cardinal axes. While the x, y and *z* values were calculated from two of the cameras, the third was used to determine appropriate conversion factors to compensate for perspective error. Duquet, Borms and Hebbelinck (1971), on the other hand, developed a tridimensional analysis technique which required only two cameras positioned at right angles to each other, one being at the side and the other overhead. If a point were visible in the films of both cameras, it could be manipulated graphically to a common plane where its location could be subsequently determined. Susanka (1969, 1970) also proposed a two-camera method for obtaining spatial coordinates.

A somewhat different approach was taken by Noble (1968), who positioned three cameras at right angles to one another along the conventional *x*, *y* and *z* axes. Horizontal and vertical coordinates were obtained from points recorded on the films of all three cameras, resulting in two *x*, two *y* and two *z* values for each point. The mean of each pair was taken to be the true coordinate. A similar method had been described earlier by Noss (1967).

In many of the multiple techniques cited, landmarks must be filmed by all cameras simultaneously if their spatial coordinates are to be derived. Because of the solid, irregular and opaque nature of the human body, the number of segmental endpoints for which *x*, *y* and *z* values can be determined by these methods is, therefore, limited. Others require the calculation of a conversion factor to compensate for perspective error. This conversion must be continually recomputed to account for changes in the subject position with relation to the cameras. In an attempt to overcome these problems, Miller and Petak (1973) developed a system in which perspective error was automatically eliminated and spatial coordinates could be determined provided a point was visible in any two of the three cameras. The underlying theory was based on the fact that the ratio of the image size to the lens-film plane distance is equal to the ratio of the object size to the lens-object distance. Figure 6-17 illustrates the geometric relationships based upon similar right triangles from which the coordinates were calculated.

An arbitrary point *P*(X,Y,Z) is photographed by Cameras One, Two and Three. The point of intersection of their optical axes, which is also the origin of the rectangular coordinate system, is designated *O*; *D* refers to the distance from the nodal plane of the camera lens to the origin; *F* represents the lens-film plane distance; and *p*(*x*,*y*) is the film image of *P* where the coordinates are taken with respect to the origin and are expressed in film dimensions. A target, carefully positioned beyond the origin and aligned with the optical axis of the camera, provides reference film coordinates of the origin (Figure 6-18).

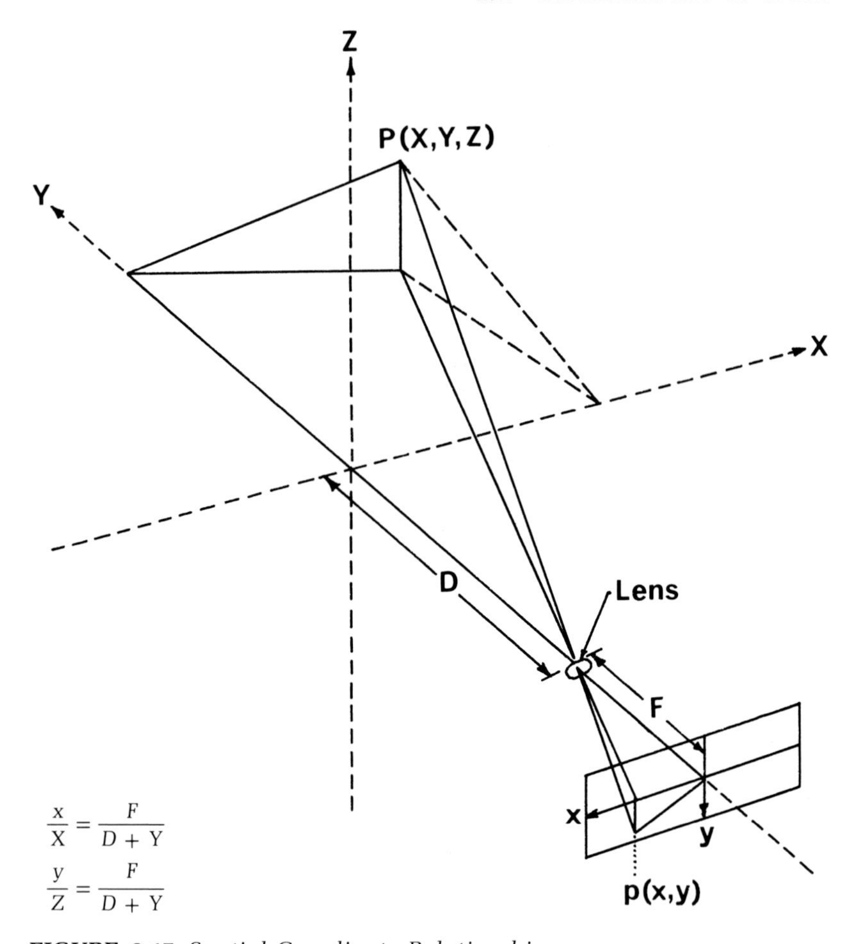

FIGURE 6-17. *Spatial Coordinate Relationships.*

From similar tetrahedrons on both sides of the lens (Figure 6-17), it can be seen that:

$$\frac{x}{X} = \frac{F}{D + Y} \qquad \text{and} \qquad \frac{y}{Z} = \frac{F}{D + Y} \ .$$

These relationships apply to Cameras One, Two and Three. Then, with capital letters and numbers indicating the real coordinates of the point as viewed by a particular camera and small letters representing the respective image coordinates, the following ratios exist:

From Camera One—
$$\frac{x1}{X1} = \frac{F1}{D1 + Y1} \qquad (1)$$

$$\frac{y1}{Z1} = \frac{F1}{D1 + Y1} \qquad (2)$$

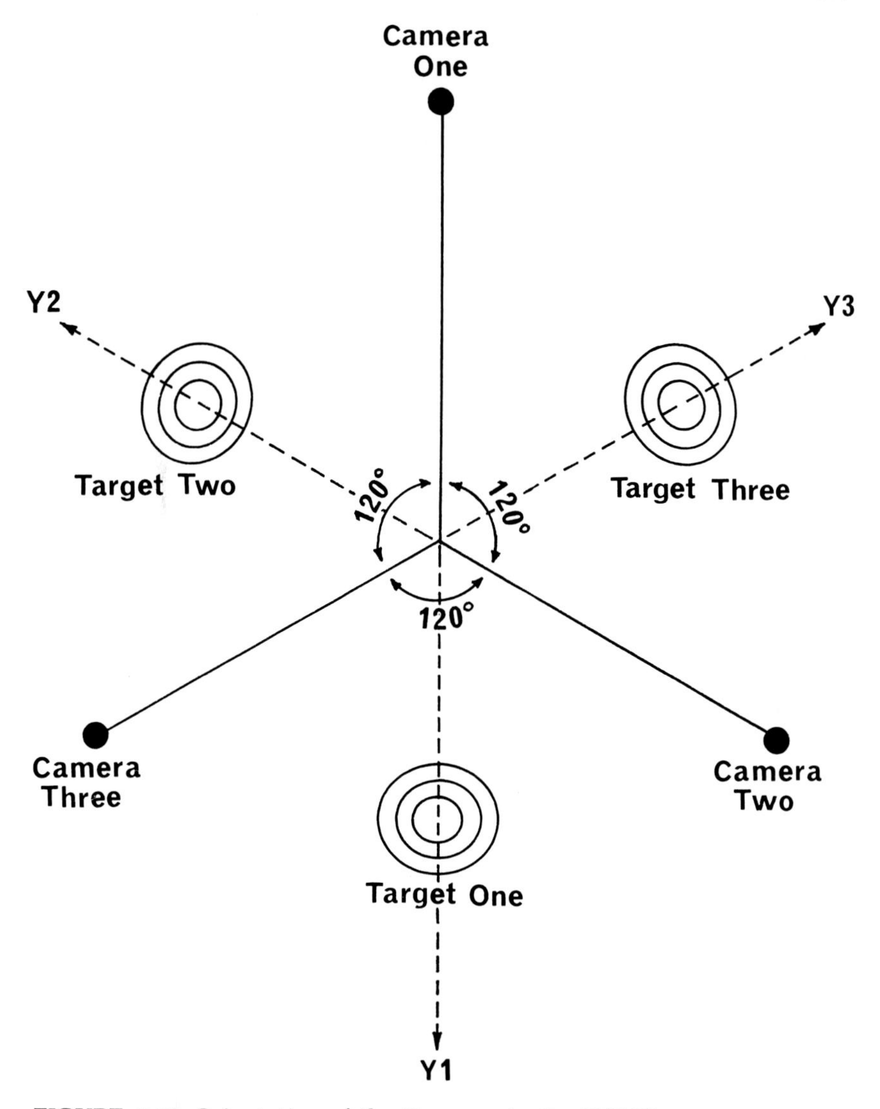

FIGURE 6-18. *Orientation of the Cameras in the X-Y Plane.*

From Camera Two— $$\frac{x2}{X2} = \frac{F2}{D2 + Y2} \qquad (3)$$

$$\frac{y2}{Z2} = \frac{F2}{D2 + Y2} \qquad (4)$$

From Camera Three— $$\frac{x3}{X3} = \frac{F3}{D3 + Y3} \qquad (5)$$

$$\frac{y3}{Z3} = \frac{F3}{D3 + Y3} \qquad (6)$$

The cameras must be positioned so that their optical axes are horizontal and intersect at a single point. The optical axes of Cameras One and Two and One and Three form 120-degree angles with one another as do those of Cameras Two and Three (Figure 6-18).† Each camera is sighted along its own positive Y axis. The coordinate system of Camera One is taken to be the main reference frame and the other two systems are rotated through 120 degrees to conform with it. Applying these transformations:

$$
\begin{aligned}
X2 &= -X1 \cos 60 - Y1 \cos 30 \\
Y2 &= X1 \cos 30 - Y1 \cos 60 \\
Z2 &= Z1 \\
X3 &= -X1 \cos 60 + Y1 \cos 30 \\
Y3 &= -X1 \cos 30 - Y1 \cos 60 \\
Z3 &= Z1
\end{aligned}
$$

Appropriate substitutions are then made for $X2$, $Y2$ and $Z2$ into equations (3) and (4) and for $X3$, $Y3$, and $Z3$ into equations (5) and (6). Thus, for any two of the three cameras, there are four equations from which to solve for the three spatial coordinates $X1$, $Y1$ and $Z1$. Although there are four different combinations of the three equations which can be formed from each pair of cameras, only two of these combinations are linearly independent. They are equations (1), (2), (3) and (1), (2), (4) from Cameras One and Two; equations (1), (2), (5) and (1), (5), (6) from Cameras One and Three; and equations (3), (4), (5) and (3), (5), (6) from Cameras Two and Three.‡

To obtain accurate results, a number of procedures must be carefully observed in the practical implementation of the method. If the study is to take place in a gymnasium or on a playing field, a surveyor's transit should be set up at the specified origin of the spatial coordinate system. The location of the cameras and their corresponding sighting targets (Figure 6-18) can then be determined following well-established surveying procedures. All cameras and targets must be on the same level. Further, the optical axes of the cameras must be horizontal and they must intersect at the origin of the coordinate system. A plumb bob suspended from the nodal plane of each lens and the center of each target; spirit levels; reflex or through-the-lens viewing; and sturdy, adjustable tripods help to achieve accurate positioning. Identification information specifying

†An angle of 120 degrees between the optical axes is assumed throughout the derivation. It is felt that this positioning of the cameras will maximize the number of points visible in the films of any two of the three cameras. Other camera separation angles are also possible provided their magnitudes are known and substituted into the appropriate places in the equations.

‡Linear independence implies that the equations have a solution which is relatively insensitive to small variations in their coefficients.

the camera, subject, trial and experimental condition as well as some type of accurate timing display must be included in the field of view of each camera. A checklist of recommended equipment and procedures is presented in Table 6-4.

Before commencing the actual study, a range pole must be held

Table 6-4. Checklist of Equipment Required for Three Dimensional Cinematography

Number	*Equipment*	*Recommendations*
3	16mm Cameras	Motor driven with the "on-off" switch remote from the camera to avoid unnecessary camera movement. If possible, all cameras should be controlled by a single switch. Through-the-lens reflex viewing is recommended. An accurately positioned reticle will also help to indicate the center of the photographic field.
3	Ciné Lenses	Good quality to minimize distortion. Appropriate focal lengths.
3	Camera Tripods	Sturdy and continuously adjustable in the horizontal and vertical directions to facilitate the fine adjustments required in camera positioning.
3	Camera Targets on Adjustable Stands	Targets should be a minimum of eight inches in diameter. Rifle targets mounted on a stiff backing will serve the purpose.
6	Plumb Bobs and String	Suspended from the nodal planes of the three camera lenses and the centers of the targets. A contrasting background should be placed behind each target plumb line so that this vertical reference can be clearly identified during film analysis.
1	Transit with Tripod	To align and level the cameras and targets. A magnifying glass, plumb bob and line should accompany the transit.
1	Level Rod with Vernier Target	To establish vertical dimensions.

Table 6-4 *Continued*

Number	*Equipment*	*Recommendations*
1	Measuring Tape	The type used in surveying which has a minimum of distortion under tension.
1	Range Pole	Symmetrical construction along its length to be used as a calibration reference.
3	Spirit Levels	One for each camera to assist in levelling.
3	Sets of Identification Numbers	Placed in one corner of the photographic field of each camera to identify camera number, subject, trial, etc. Lap counters used in swimming or track are ideal for this purpose.
2,3	Timing Displays	Digital readout indicating time to .001 of a second with one being visible in the field of each camera. Multiple timing displays must be controlled by a single pulse generator. Mirrors may be useful.
	Artificial Lighting	If required.
	Power Outlets	For cameras, timing displays, lights.
	Extension Cords	

in a vertical position at the origin and filmed by all cameras. Since the contrasting sections of the pole are of known length, they provide a ready means for calculating the lens-to-film plane distance. The latter is usually slightly larger than the focal length of the lens and must be determined from the relationship:

$$\frac{\text{Lens-to-film distance}}{\text{Image size}} = \frac{\text{Lens-to-object distance}}{\text{Object size}}.$$

The image of the range pole itself or a selected portion of it must be measured on a film analyzer and subsequently divided by the projector magnification to reduce it to its actual size on the film. Substitution of the appropriate values into the ratio then yields the lens-to-film distance.

When obtaining data from the films, a motion analyzer of good quality should be used. Frames exposed at corresponding instants of time have to be matched with the aid of the timing display visible in the field of each camera. Correct vertical alignment of the selected frames is then obtained by using the plumb line suspended from the center of the sighting target which appears in the background of the picture. The coordinates of the center of this target, which represent the origin of the system, must be recorded along with the x and y coordinates of specific points on the film image. Provided that a point is visible in any two of the cameras simultaneously, its spatial coordinates can be determined.

Because of the extensive calculations required, the raw data must be processed by digital computer. A program written for this purpose would subtract the x and y coordinates of the target center from those of designated points on the film image. The resulting values would then be divided by the analyzer magnification to yield coordinates expressed in film dimensions with respect to the origin of the system. Input to the computer program would include the distances from the cameras to the origin as well as the necessary information to calculate their lens-to-film plane distances. Substitution of these data into the appropriate equations would produce the desired X, Y and Z coordinates.

The three-dimensional method outlined is more flexible than those proposed previously, and therefore, should have wider application. The cameras can be positioned arbitrarily provided that their optical axes are horizontal and intersect at a common point which serves as the origin of the coordinate system. Lenses of any suitable focal length and distance setting may be used. If a data point is visible in any two cameras, its x, y and z coordinates can be determined. Since the equations automatically compensate for movement of the subject toward or away from the cameras, there is no need to constantly recalculate conversion factors.

SUMMARY

Although a number of methods have been suggested for obtaining spatial coordinates from film, few have been used extensively in research applications. All require a considerable expenditure of time and effort in preparing the filming site for data collection. Even then, it may be difficult to obtain sufficient information to calculate x, y and z coordinates for all the desired body landmarks. Extracting the raw position data from the film is more laborious than with planar methods. In addition, one of the major problems with using two or more cameras is the identification of simultaneously exposed frames. In some instances, interpolation techniques may be required to determine coordinates at specified instants in time. Because of the lengthy calculations involved in three-dimensional cinematography, the use of the digital computer for data reduction and processing is almost mandatory. Regardless of which method is selected, it is the investigator's responsibility to verify the experimental protocol, equations and computer routines by determining the spatial coordinates of an object of known position and dimensions. Such calculations will help identify gross errors and will provide an indication of the relative accuracy of the technique.

SELECTED REFERENCES

Anderson, C. C.: A Method of Data Collection and Processing for Cinematographic Analysis of Human Movement in Three Dimensions. Unpublished Master's Thesis, University of Wisconsin, 1970.

Atwater, A. E.: University of Arizona, Personal Communication, 1972.

Atwater, A. E.: Movement Characteristics of the Overarm Throw: A Kinematic Analysis of Men and Women Performers. Unpublished Doctoral Dissertation, University of Wisconsin, 1970.

Ayoub, M. A., Ayoub, M. M., and Ramsey, J. D.: A Stereometric System for Measuring Human Motion. Hum. Factors, *12*, 523–535, 1970.

Barlow, D. A.: Relationship between Power and Selected Variables in the Vertical Jump. Unpublished Master's Thesis, Pennsylvania State University, 1970.

Baumann, W.: Über die kinematografische Bewegungsanalyse. Med. Welt, *19*,2168–2174, 1968.

Bernstein, N.: *The Co-ordination and Regulation of Movements*. New York: Pergamon Press, 1967.

Bernstein, N.: Untersuchung der Korperbewegungen und Korperstellungen im Raum mittels Spiegelaufnahmen. Int. Z. Angew. Physiol., *3*, 179–206, 1930.

Blievernicht, D. L.: A Multidimensional Timing Device for Cinematography. Res. Q. Amer. Assoc. Health Phys. Ed., *38*, 146–148, 1967.

Burton, A. L. (Ed.): *Cinematographic Techniques in Biology and Medicine*. New York: Academic Press, 1971.

Cimerman, V. J., and Tomasegovic, Z.: *Atlas of Photogrammetric Instruments*. New York: American Elsevier, 1970.

Cooper, J. M., and Sorani, R. P.: Use of the Dichroic Mirror as a Cinematographic Aid in the Study of Human Performance. Res. Q. Amer. Assoc. Health Phys. Ed., *36*, 210–211, 1965.

Cureton, T. K.: Elementary Principles and Techniques of Cinematographic Analysis as Aids in Athletic Research. Res. Q. Amer. Assoc. Health Phys. Ed., *10*, 3–24, 1939.

Davies, E. R.: Photographic Analysis of Motion. Nature, *152*, 261–264, 1943.

Doolittle, T. L.: Errors in Linear Measurement with Cinematographical Analysis. In *Kinesiology* 1971, Washington: AAHPER, *1971*.

Drillis, R.: The Use of Gliding Cyclograms in the Biomechanical Analysis of Movements. Hum. Factors (April), *1*, 1–11, 1959.

Drillis, R.: Objective Recording and Biomechanics of Pathalogical Gait. Ann. N. Y. Acad. Sci., *74*, 86–109, 1958.

Drillis, R.: Photographic Evaluation of Prosthetics. Indust. Photogr., *5*, 24–25, 1956.

Dubovik, A. S.: *Photographic Recording of High-speed Processes*. New York: Pergamon Press, 1968.

Duquet, W., Borms, J., and Hebbelinck, M.: A Method of Tridimensional Analysis of Twisting Movements. Paper presented at the Third International Seminar on Biomechanics, Rome, 1971.

Eberhart, H. D., and Inman, V. T.: An Evaluation of Experimental Procedures Used in a Fundamental Study of Human Locomotion. Ann. N. Y. Acad. Sci., *51*, 1213–1228, 1951.

Edgerton, H. E., Germeshausen, J. K., and Grier, H. E.: High Speed Photographic Methods of Measurement. J. Appl. Phys., *8*, 2–9, 1937.

Engel, C. E. (Ed.): *Photography for the Scientist*. New York: Academic Press, 1968.

Focal Encyclopedia of Photography (Rev. Desk Ed.). New York: McGraw-Hill, 1969.

Garnov, V. V., and Dubovik, A. S.: Stereoscopic Filming of Rapid Processes by Two Independently Operating Moving Picture Cameras. 1965 (NASA TT F-337).

Gavan, J. A., Washburn, S. L., and Lewis, P. H.: Photography: An Anthropometric Tool. Amer. J. Phys. Anthropol., *10*, 331–353, 1952.

Groh, H., and Baumann, W.: Kinematische Bewegungsanalyse. In J. Wartenweiler *et al.* (Eds.), *Biomechanics*. Basel: Karger, 1968.

Gutewort, W.: The Numerical Presentation of the Kinematics of Human Body Motions. In J. Vregenbregt and J. Wartenweiler (Eds.), *Biomechanics II*. Basel: Karger, 1971.

Gutewort, W.: Die digitale Erfassung kinematischer Parameter der menschlichen Bewegung. In J. Wartenweiler *et al.* (Eds.), *Biomechanics*. Basel: Karger, 1968.

Hallert, B.: *Photogrammetry*. New York: McGraw-Hill, 1960.

Hellebrandt, F. A., Hellebrandt, E. T. and White, C. H.: Methods of Recording Movement—An Appraisal of Photographic Techniques for the Study of Motor Performance and Learning. Amer. J. Phys. Med., *39*: 178–183, 1960.

Herron, R. E.: Stereophotogrammetry in Biology and Medicine. Photogr. Appl. Sci. Technol. Med., *5*, 26–35, 1970 (September).

Herron, R. E.: A Biomedical Perspective on Stereographic Anthropometry. In F. D. Thomas and E. Sellers (Eds.), *Biomedical Instrumentation*. Vol. 6. Pittsburgh: Instrument Society of America, 1969.

Hubbard, A. W.: Photography. In M. G. Scott (Ed.), *Research Methods in Health, Physical Education and Recreation*. Washington: AAHPER, 1959.

Hyzer, W. G.: How Accurate Are Photographic Measurements? Res. Develop., *22*, 43–44, 1971a (August).

Hyzer, W. G.: How Accurate are Photographic Measurements? Part II. Res. Develop., *22*, 75–78, 1971b (October).

Hyzer, W. G.: Camera Systems for R and D. Res. Develop., *22*, 63–68, 1971c (April).

Hyzer, W. G.: *Engineering and Scientific High-Speed Photography.* New York: Macmillan, 1962.

Jones, F. P., and O'Connell, D. N.: Color-Coding of Stroboscopic Multiple-Image Photographs. Sci., *127,* 1119, 1958.

Jones, F. P., and O'Connell, D. N.: Applications of Multiple-Image Photography in the Time-Motion Analysis of Human Movement. Photogr. Sci. Technol., *3,* 11–14, 1956.

Jones, F. P., O'Connell, D. N., and Hanson, J. A.: Color-Coded Multiple-Image Photography for Studying Related Rates of Movement. J. Psychol., *45:* 247–251, 1958.

Kasvand, T.: Pattern Recognition Applied to Measurement of Human Limb Positions during Movement. Division of Mechanical Engineering, National Research Council, Ottawa, 1970 (LTR-CS-41).

Kasvand, T., Milner, M., and Rapley, L. F.: A Computer-Based System for the Analysis of Some Aspects of Human Locomotion. Proceedings of the Conference on Human Locomotor Engineering. University of Sussex, 1971.

Ketlinski, R.: Can High Speed Photography be Used as a Tool in Biomechanics? In J. M. Cooper (Ed.), *Selected Topics on Biomechanics.* Chicago: Athletic Institute, 1971.

Kissam, P.: *Optical Tooling for Precise Manufacture and Alignment.* New York: McGraw-Hill, 1962.

Kodak Customer Service Pamphlets. Rochester: Eastman Kodak.

Korn, J. (Ed.): *Life Library of Photography.* New York: Time-Life Books, 1970.

Lamaster, M. A., and Mortimer, E. M.: A Device to Measure Body Rotation in Film Analysis. Paper presented at the AAHPER Convention, Washington, D.C., 1964.

Marey, J.: The History of Chronophotography. In the Smithsonian Institute Annual Report for 1901. Washington: Government Printing Office, 1902.

Miller, D. I., and Petak, K. L.: Three Dimensional Cinematography. In *Kinesiology 1973.* Washington: AAHPER, 1973.

Moffitt, F. H. *Photogrammetry.* 2nd Ed., Scranton, Pa.: International Textbook Company, 1967.

Muybridge, E.: *Animals in Motion.* New York: Dover, 1957.

Muybridge, E.: *The Human Figure in Motion.* New York: Dover, 1955.

Nelson, R. C., and Miller, D. I.: Cinematography in Biomechanics Research. Res. Film, *7,* 95–102, 1970.

Nelson, R. C., Petak, K. L., and Pechar, G. S.: Use of Stroboscopic-Photographic Techniques in Biomechanics Research. Res. Q. Amer. Assoc. Health Phys. Ed., *40,* 424–426, 1969.

Noble, M. L.: Accuracy of Tri-Axial Cinematographic Analysis in Determining Parameters of Curvilinear Motion. Unpublished Master's Thesis, University of Maryland, 1968.

Noble, M. L., and Kelley, D. L.: Accuracy of Tri-Axial Cinematographic Analysis in Determining Parameters of Curvilinear Motion. Res. Q. Amer. Assoc. Health Phys. Ed., *40,* 643–645, 1969.

Noss, J.: Control of Photographic Perspective in Motion Analysis. JOHPER, *38,* 81–84, 1967 (September).

Novac, A.: L'Utilisation des Techniques Cinematographiques pour l'Etude du Mouvement Sportif, II. Educ. Phys. Sport, *78*, 15–18, 1966 (January).

Novac, A.: L'Utilisation des Techniques Cinematographiques pour l'Etude du Mouvement Sportif, I. Educ. Phys. Sport, *77*, 15–18, 1965 (November).

Plagenhoef, S.: Computer Programs for Obtaining Kinetic Data on Human Movement. J. Biomech., *1*, 221–234, 1968.

Quanbury, A. O.: National Research Council, Ottawa. Personal Communication, 1972.

Richards, C. L.: Computer Analysis of a Cinematographic Study of Normal Natural Gait in Children. Unpublished Master's Thesis, University of Saskatchewan, 1972.

Roberts, E. M.: Cinematography in Biomechanical Investigation. In J. M. Cooper (Ed.), *Selected Topics on Biomechanics*. Chicago: Athletic Institute, 1971.

Susanka, P.: Nektere Aspekty Kinematicke Analyzy Pohybu (Se Zamerenim Na Doinf Koncetiny). Katedra Antropomotoriky a Biomechaniky, Charles University, Prague, 1970.

Susanka, P., and Diblik, J.: Raumkinematographie. In the Proceedings of the Second International Seminar on Biomechanics, Eindhoven, 1969.

Taylor, P. R.: Essentials in Cinematographical Analysis. In J. M. Cooper (Ed.), *Selected Topics on Biomechanics*. Chicago: Athletic Institute, 1971.

Time, Inc.: Buyer's Guide to Camera Accessories. New York: Time-Life Books, 1970.

Walton, J. S.: Photographic and Computation Techniques for Three-Dimensional Location of Trampolinists. Unpublished Master's Thesis, Michigan State University, 1970.

Walton, J. S.: A High Speed Timing Unit for Cinematography. Res. Q. Amer. Assoc. Health Phys. Ed., *41*, 213–216, 1970.

CHAPTER 7

Electronic Instrumentation

APPLICATION OF electronic instrumentation to research in biomechanics of sport has occurred concurrently with the use of photoinstrumentation techniques. Developments in the field of electronics in the past two decades now make it possible to measure and record a variety of physical phenomena associated with human motion. Although it is necessary for the subjects to be in contact with at least the sensing component of the system, data collected by this direct measurement may be more accurate than that attainable with other methods. Because electronics has become essential to the study of biomechanics, researchers must develop a working knowledge of its fundamentals and the types of measurement systems available. This knowledge will permit the researcher to communicate with sales representatives and to interact with a technically qualified staff member. As will be noted in Chapter 11, it has become imperative that a technical assistant be available to work closely with the researcher.

A complete discussion of the field of electronics is beyond the scope of this text.† Rather, a general introduction to timers and instrumentation systems including the most commonly used transducers, signal conditioners and recorders is presented. The following definitions of terms are included to acquaint the reader with essential terminology. They represent a portion of a more complete glossary found in the Honeywell Instrumentation Handbook (1970, pp. 18–27).

DEFINITIONS

AC transducer: a transducer which must be excited with alternating current in order to operate properly and whose output appears in the form of an alternating current.

Amplitude: a measure of the deviation of a phenomenon from its average or mean position.

Analog output: transducer output which is a continuous function of the measurand except as modified by the resolution of the transducer.

Attenuation: the reduction in amplitude of a given stimulus or signal.

Bidirectional transducer: a device capable of measuring stimuli in both a positive and a negative direction from a zero position.

Binary coded decimal: a type of display in which the numerical information is presented in decimal form but each decimal digit is represented by a code of four binary digits.

Bonded strain gage: strain-sensitive elements arranged to facilitate bonding to a surface in order to measure applied stresses.

Bridge: a term denoting the general electrical configuration of certain transduction elements which also serves as an abbreviation of "Wheatstone Bridge."

Calibration: the known values of the measurand are applied while the transducer output is observed or recorded. The calibration data provide information pertaining to the non-linearity, combined non-linearity and hysteresis, and/or hysteresis characteristics of the transducer.

Calibration curve: graphic presentation of data obtained during calibration.

Compensation: capability of a device to counteract known sources of error.

Crystal transducer: a transducer in which the transduction is accomplished by means of the piezoelectric properties of certain crystals or salts.

† The reader is referred to Alnutt and Weinberg (1963) and Alnutt and Becker (1964), Brophy (1966), Norton (1969) and the Honeywell Instrumentation Handbook (1970) for background information.

Current: flow of electrons caused by a difference in potential between two points.

DC transducer: a transducer which operates when excited with direct current and whose output is normally direct current.

Damped natural frequency: the frequency of uninhibited oscillations of the sensing element of a transducer.

Damping: the energy-dissipating characteristic which, together with natural frequency, determines the upper limit of frequency response and the response-time characteristics of a transducer.

Digital output: transducer output that represents the magnitude of the measurand in the form of a series of discrete quantities coded in a system of notation.

Digitizer: a device which converts analog data into digital form.

Error: the algebraic difference between the indicated value and the true value of the measurand expressed in percent of the full-scale output or in percent of the output reading of the transducer.

Excitation: the external electrical voltage and/or current applied to a transducer for its proper operation.

Frequency response: the change with frequency of the output/measurand amplitude ratio for a sinusoidally varying measurand applied to a transducer within a stated range of measurand frequencies.

Full scale: total stimulus interval over which the instrument is intended to operate.

Gage: an instrument or means for measuring or testing. This term is often used synonymously with "transducer."

Hertz: term used to indicate cycles per second.

Hysteresis: the maximum difference in output at any given measurand value within the specified range when the value is approached first with increasing and then with decreasing measurand.

Inductive transducer: a transducer which receives stimulus information by means of changes in inductance.

Integrating accelerometer: a transducer designed to measure velocity by means of a time integration of acceleration.

Linearity: the closeness of a calibration curve to a specified straight line.

Measurand: a physical quantity, property or condition which is measured.

Multiplexing: the simultaneous transmission of two or more signals within a single channel. The three basic methods of multiplexing involve the separation of signals by time division, frequency division or phase division.

Natural frequency: see definition of "damped natural frequency."

Noise: any unwanted electrical disturbance or spurious signal which modifies transmitting, displaying or recording of desired data.

Null: a condition of balance which results in a minimum absolute value of output.

Overload: the amount beyond the specified maximum magnitude of the measurand which, when applied to a transducer, does not cause a change in performance beyond specified tolerance.

Peak-to-peak: the algebraic difference between maximum positive and negative values of a varying stimulus or signal.

Pot: an abbreviation for potentiometer.

Potentiometric transducer: a transducer in which the displacement of the force-summing member is transmitted to the slider in a potentiometer, thus changing the ratio of output resistance to total resistance.

Power: energy transferred per unit of time. Work per unit of time.

Range: the spectrum of measurand values which exists between the upper and lower limits of the transducer's measuring capability.

Readout equipment: the electronic apparatus employed to provide indications and/or recordings of a transducer output.

Reliability: a measure of the probability that an instrument will continue to perform within specified limits of error for a specified length of time and under specified conditions.

Repeatability: the ability of a transducer to indicate consistently the same output readings for given measurand values.

Resolution: magnitude of output step changes (expressed in percent of full-scale output) as the measurand is varied continuously over the range.

Scale factor: the ratio of full-scale output to the value of the measurand at full range.

Seismic mass: the element in an accelerometer intended to serve as the force-summing member for applied accelerations and/or gravitational forces.

Sensitivity: the ratio of the change in transducer output to a change in the value of the measurand.

Sensitivity drift: a slow, continuous modification in sensitivity due to a variety of internal causes.

Signal: the output emanating from a device.

Stability: the ability of a transducer to retain its performance capability.

Strain: elastic deformation produced in a solid as a result of stress.

Strain gage: a device for converting the measurand into a change of resistance due to strain, usually in two or four legs of a Wheatstone Bridge.

Stress: the force acting on a unit area of a solid.

Telemetry: a measurement accomplished with the aid of intermediate means which allow perception, recording or interpretation of data at a distance from a primary sensor.

Time base: a reference time signal recorded at given intervals or continuously with the information signal.

Transducer: broadly defined, a device which enables energy to flow from one or more transmission systems to one or more other transmission systems. The term "transducer" is most often restricted to signify a device in which the magnitude of an applied stimulus is converted into an electrical signal which is proportional to the quantity of the stimulus.

Undamped natural frequency: the natural frequency if no damping is present.

Velocity transducer: a transducer which produces an output proportional to imparted velocities.

Voltage: the potential difference between two points. If no current flows it is referred to as electromotive force.

Zero adjustment: the act of nulling the output from a system or device.

Zero shift: an error characterized by a parallel displacement of the entire calibration curve.

Timing Devices

As noted in Chapter 4, biomechanical analysis usually begins with quantification of the temporal components of the movement under investigation. In many instances cinematography provides a very satisfactory method for measurement of time. The fact that the results are not readily available, however, may prove to be a serious limitation in some research applications. Consequently, a variety of timing devices has been used to measure time duration of selected phases of human movements. These timers rely either on the standard 60 Hz frequency or high-frequency oscillators to provide an accurate time base. The timer may be triggered to start and stop with microswitches, photocells, pressure switches or magnetic switches.

Such a system was used by Henry and Whitley (1960) to determine the execution time of the adductive arm swing. A microswitch depressed by the subject prior to movement started a chronoscope which was later stopped as the subject's arm contacted a string. A similar apparatus for measurement of elbow flexion speed was developed by Nelson and Fahrney (1965), who utilized photo-cell circuitry to start and stop an electronic counter. In an attempt to measure movement time through segments of an arm motion, Henry (1960) devised an apparatus with a series of strings along the arm swing path which, when dislodged, stopped one timer and started the next. A unit in which small wooden dowels replaced the strings as the triggering mechanism was reported by Whitley and Smith (1963). To measure the velocity of a thrown ball, Nelson and associates (1966) utilized two photosensitive fields positioned four feet apart. As the ball passed through the first field a Beckman Timer was started and subsequently stopped as the ball continued through the second field.

A timing unit called the Automatic Performance Analyzer (APA) (manufactured by Dekan Timing Devices) meets a wide range of practical timing requirements. This portable unit which measures time to 0.01 sec. can be started and stopped in a number of ways including control cord, switch mat, push button and external switch. Starting signals include a buzzer, delayed buzzer and lamp. The APA can be used to measure simple reaction and movement times as well as more complex movements in sport. For example, the 10 and 40 yard times of Penn State University football players were determined with the unit displayed in Figure 7-1. The players started the timer by releasing a hand switch and, after running the prescribed distance, stopped it by striking the string which removed the attached plug from the stop switch. A similar system was used by Scheuchenzuber (1970) in his study of reaction and movement time in the backstroke start in swimming. An electronic counter was triggered as the gun

FIGURE 7-1. *Use of the Automatic Performance Analyzer to Time Football Players.*

sounded and stopped as the swimmer's hands released the switch on the starting bar (Figure 7-2).

A versatile, accurate timing device is essential to a research program in biomechanics. For example, the Hickok digital counter shown in Figure 7-3 can be successfully used in a variety of applications. In this instance it was used by Brooks (1973) to accurately measure the period of a swinging pendulum which was required in her calculation of the moment of inertia of the leg of lamb. A small photocell positioned at the low point of the swing provided the means for triggering the counter.

These relatively inexpensive timing devices form the basis of instrumentation for biomechanics research. In many situations, initial purchase of timing equipment represents the most economical choice since the equipment can be used extensively in teaching and research. The examples presented here are but a few of the possible applications.

FIGURE 7-2. *Special Apparatus for Measuring Reaction Time in the Backstroke Start (From Scheuchenzuber, H. J.: A Biomechanical Analysis of Four Backstroke Starts. Unpublished Master's Thesis, Pennsylvania State University, 1970).*

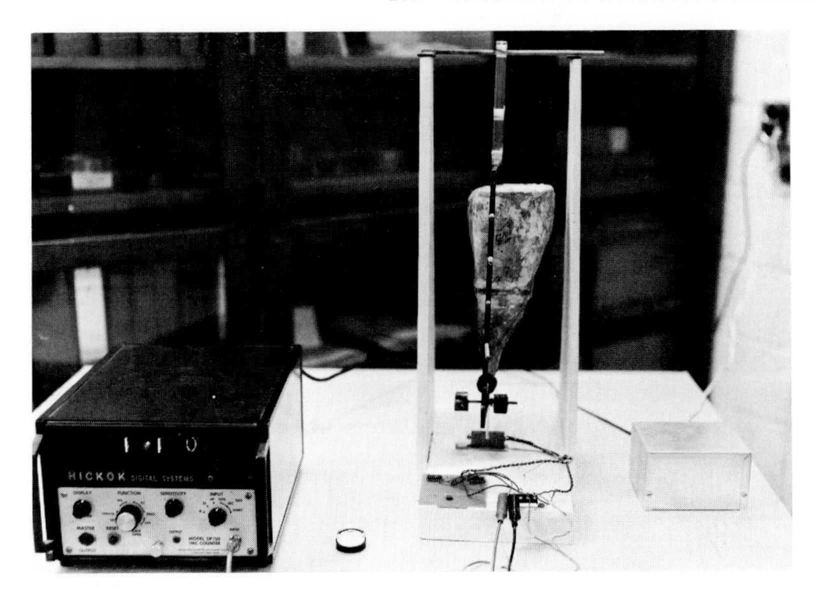

FIGURE 7-3. *Digital Counter Used to Measure Pendulum Period (From Brooks, C. M.: Validation of the Gamma Mass Scanner for Determination of Center of Gravity and Moment of Inertia of Biological Tissue. Unpublished Master's Thesis, Pennsylvania State University, 1973).*

INSTRUMENTATION SYSTEMS

In its simplest form, an instrumentation system consists of a transducer, signal conditioner, power supply and recorder. The transducer, which serves as a sensing device, either generates or modifies a signal by an amount related to the magnitude of the variable under study. The signal conditioner converts the output from the transducer into an electrical quantity which is compatible with the display and/or recording device. The power supply provides excitation for the transducer, power for the signal conditioner and in some cases for the display device. The latter permits observation of the phenomena being measured while the recording unit provides a permanent record. In some systems, the components may not be physically separated. For example, the signal conditioner and/or power supply may be incorporated in either transducer or readout unit (Norton, 1969).

Transducers. The most commonly used transducers in biomechanics research are the mechanical and electronic types. The mechanical devices include the spring dynamometer and the cable tensiometer. These relatively simple instruments have been used successfully to measure maximum human forces of an isometric nature. However, recent research has shifted from merely assessing maximum force to an examination of the interaction of force and time during muscular contraction, as noted in Chapter 3. For this reason and because of the interest in a variety of human performance parameters, electronic transducers have become the primary sensing devices used in biomechanics measurement systems. Although a large variety of such transducers is available (see Norton, 1969), only those most frequently used in biomechanics are included here. These are the potentiometer, strain gage, linear variable differential transformer (LVDT) and accelerometer.

Potentiometric devices are capable of sensing linear as well as angular position. However, the latter type are of considerably greater importance in the study of human limb movements and, as a consequence, have been more widely used. These relatively simple, inexpensive devices which function as variable resistance transducers are composed of a sliding contact (wiper) which moves over a resistance (potentiometric) element. Angular motion of the slider arm modifies the DC voltage, which is then amplified and recorded (Norton, 1969). Rotation of a few degrees to a complete revolution or more is possible. Since the joints of the body permit less than 360 degrees of rotation (most less than 180 degrees), it may be advisable to use a "sector" potentiometer which will maximize the sensitivity through the range of motion being studied. The frequency response which is determined by the mechanical linkage associated with the transducer and the mass of the wiper arm (Alnutt and Weinberg, 1963) is normally less than 100 Hz, which is adequate for monitoring human limb movements. Serious problems will arise, however, if the output of the potentiometer is not linear.

Continuous recording of angular position versus time provides important information about the characteristics of joint motion and also permits further study of the kinematics and dynamic phenomena of the movement (Koniar, 1968). The potentiometer has been utilized primarily in two ways for this purpose. In the first technique, the device is secured to the test apparatus at the point of rotation. Such a device was used by Stothart (1970) to record position during an elbow flexion movement (Figure 7-4).

The second and most common use of the potentiometer has been as a component in the electrogoniometer (elgon) developed originally by Karpovich (1959). Details of the construction, calibration and operational procedures for this device have been reported by a number of authors (Karpovich *et al.*, 1960; Adrian, 1966; Adrian, 1968; Hutinger, 1971; and Korb, 1970). The elgon is positioned with the potentiometer over the joint selected for motion study (Figure 7-5).

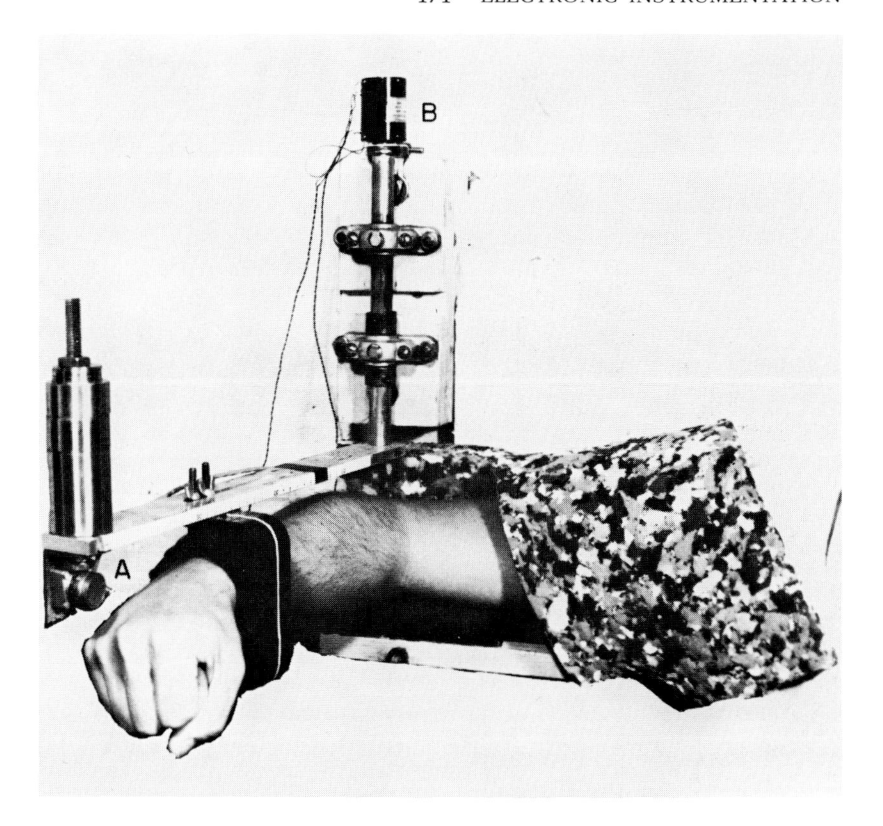

FIGURE 7-4. *Axle-mounted Potentiometer for Measurement of Angular Displacement. A) Accelerometer B) Potentiometer (From Stothart, J. P.: A Biomechanical Analysis of Static and Dynamic Muscular Contraction. Unpublished Doctoral Dissertation, Pennsylvania State University, 1970).*

The arms of the unit are strapped to the limbs adjacent to the joint and move with them.

The elgon offers one means of recording continuous action of selected joints of the body and has been used successfully to study a variety of sports skills. Gollnick and Karpovich (1964) reported results obtained from knee and ankle goniograms of subjects performing the back handspring, front and back somersaults, selected swimming strokes, a baseball throw and a shot put. Results of electrogoniometric studies of swimming strokes, running and shot putting were presented by Koniar (1968) while Ringer and Adrian (1969) investigated the motion at the wrist and elbow in the crawl arm stroke. The high jump, broad jump and triple jump were the focus of research reported by Klissouras and Karpovich (1967). Practical application of this technique to improve performance has

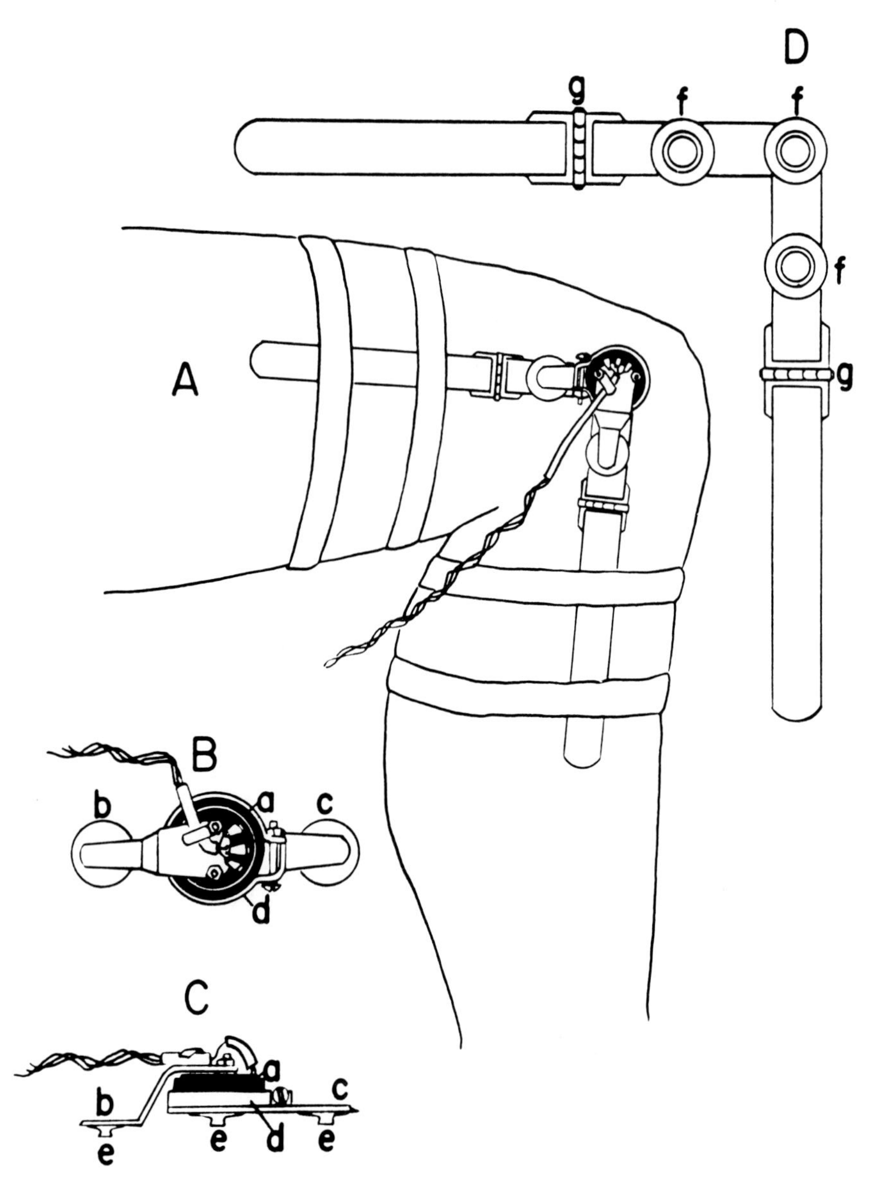

FIGURE 7-5. *Elgon Positioned for Monitoring Knee Joint Motion (From Karpovich, P. V., and Sinning, W. E.: Physiology of Muscular Activity. Philadelphia: W. B. Saunders, 1971).*

been reported by Koniar (1968), who indicated that the electrogoniogram and the derived kinematic variables of selected joints of the body represent the structure of the movement. Such an analysis serves as an evaluative measure of changes occurring during training and provides for comparison among performers of different abilities. Further, the coordination of body segments and their influence upon the total movement of the body can be ascertained.

Several limitations are associated with the use of electrogoniometer instrumentation. The attachment of the elgon to the limbs of the subject may alter normal movement patterns. The wire accessories cause further complications and require that provision be made to transport the wires if the subject is required to move more than a few feet. Further, the analysis must normally be restricted to a limited number of joints whose motion is in a single plane. An attempt has been made by Johnson and Smidt (1969) to overcome this problem by using three electrogoniometers to monitor the multi-dimensional movements of the hip joint. The mere recording of angular motion is not sufficient to study accurately the movement of the total body, and consequently, electrogoniometry should be supplemented with cinematographic or force recording methods. Use of the elgon implies a single, fixed center of rotation which is often not the case in the movement of human joints.

Investigators have attempted to develop new techniques for measuring angular displacement which might overcome some of the limitations of cinematography and electrogoniometry. One approach has been through use of polarized light methods as reported by Mitchelson (1971), Reed and Reynolds (1969) and Grieve (1969). Working independently these investigators developed similar instrument systems for recording angular position of human limbs during movement. The unit (termed the polgon by Grieve) is composed of an optical projector and a selected number of receivers attached to the body parts under investigation. The light emitted from the projector is directed at the subject through a polarized screen rotating at a predetermined speed. Outputs from the two photocells are wired in opposition so that non-polarized light incident to them will be rejected. The signal produced by the rotating plane of polarized light is sinusoidal, with a frequency twice the rate of the rotating disc (Mitchelson, 1971). The elapsed time between the reference pulse from the rotating screen and the zero point of the sinusoidal signal is electronically corrected to a DC output voltage which is converted to angular displacement from the reference position. Additional circuitry makes possible the measurement of angular velocity and acceleration. Preliminary studies have utilized paper chart recording methods but on-line recording by computer is also possible (Mitchelson, 1971).

Although these polarized light techniques have not as yet been widely used in research applications, it appears that this approach may become a significant addition to the methods of measurement of human movement parameters.

The strain gage transducer has also proven to be an effective means of measuring selected biomechanical parameters. Three types, electrolytic, semiconductor and wire, are available (Alnutt and Weinberg, 1963), but the latter is most commonly used in biomechanics research. The wire strain gage can be either bonded or unbonded. In both cases deformation, which is directly related to force, is measured by observing the change in resistance of a wire segment positioned over the moving or stressed member (Alnutt and Weinberg, 1963; Brendel, 1969). For example, a strain gage unit may be positioned in a strength testing apparatus so that it responds to the force being applied. The altered resistance due to the strain results in changes in the output voltage which are monitored and recorded. Calibration procedures provide direct conversion values from voltage to pounds or kilograms. Since any change in the temperature of the strain gage wire affects its resistance, force measurements are subject to error if the environmental temperature changes (Korn and Simpson, 1970). This problem may be overcome by temperature-compensating circuitry or frequent calibration.

The linear variable differential transformer (LVDT) has been used effectively for measurement of force. The unit is composed of a reluctance (differential-transformer) force transducer mounted inside a proving ring (Figure 7-6). Application of force to the ring causes relative motion between a core and a concentric coil assembly. Displacement of the core causes a change in amplitude and phase of the AC voltage in two secondary windings which is directly related to the applied force (Norton, 1969). Units are available with frequency responses ranging from 50 to 10,000 Hz, with maximum forces to cover the complete range which can be produced by human subjects.

The LVDT has been employed to measure human strength in biomechanics research. Figure 7-7 depicts an apparatus used by Stothart (1970) in his study of the static and dynamic characteristics of a simple movement. A similar application is shown in Figure 7-8, in which the influence of elbow angle on the force of an isometric contraction can be investigated. As the subject pulls on the apparatus, the force is transmitted through the axle-lever arm system to the LVDT force ring. The force exerted at the cuff can be calculated from the recorded force and the lengths of the two lever arms.

The importance of muscular force in the execution of sports skills and human movement in general is widely acknowledged. The research literature contains numerous studies concerned with strength development and comparisons of various training systems. Less attention has been given to investigation of the effects of muscle force upon movement characteristics or the interacting forces be-

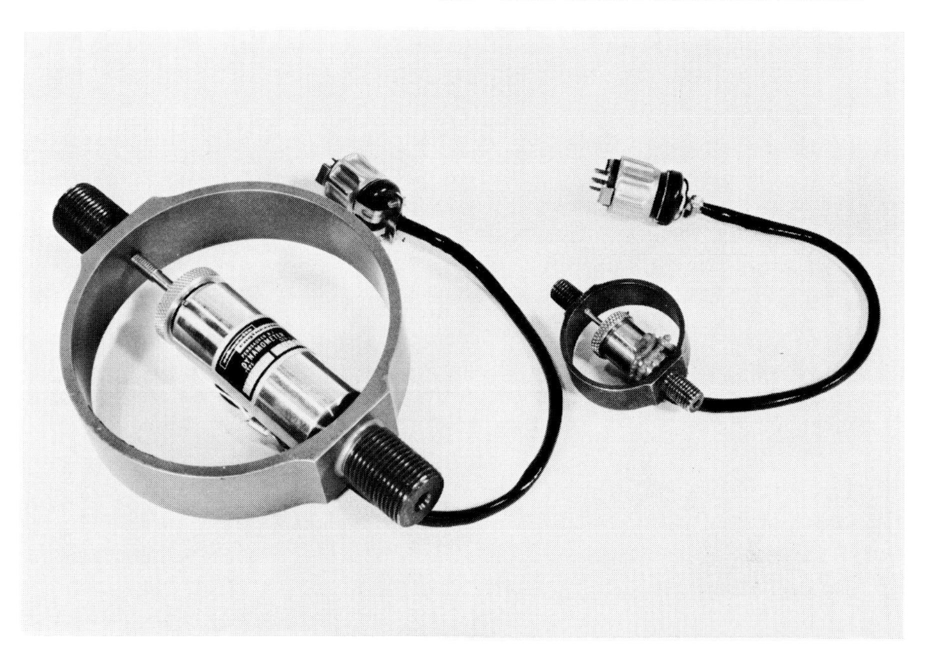

FIGURE 7-6. *LVDTs Mounted in Proving Rings.*

FIGURE 7-7. *A Depicts LVDT Used in Elbow Flexion Force Measurement (From Stothart, J. P.: A Biomechanical Analysis of Static and Dynamic Muscular Contraction. Unpublished Doctoral Dissertation, Pennsylvania State University, 1970).*

FIGURE 7-8. *Strength-testing System Utilizing LVDT.*

tween a sportsman and his sports implement or his environment (e.g., the ground, springboard, etc.). Measurement of such forces during execution of sports skills provides valuable biomechanical data concerning their performance. To investigate these forces researchers have utilized a variety of sensing platforms coupled with appropriate amplification and recording devices. The simplest systems merely determine vertical forces while more sophisticated ones measure forces in the X, Y and Z directions and the moments about each of these axes. The unit consists of a contact area constructed from layers of rigid materials and force-sensing instruments mounted in the apparatus.

Greene and Morris (1958) utilized a triangle-shaped force platform to evaluate movements in an industrial setting. The sensing devices were three LVDTs which made possible the measurement of forces in the X, Y and Z planes. A similar device, termed a power plate, in which 32 strain gages were utilized was described by Carlsöö (1962). Recording of force in the X, Y and Z directions and torques about these axes was accomplished. Recently developed platforms have been described by Payne and his associates (1968) and Hearn and Konz (1968). The latter unit, utilizing LVDT-sensing devices, provided for measurement of force in the three cardinal planes and the torque about each of three orthogonal axes. According to the authors, it provided greater sensitivity and improved repeatability.

Similar measurements were obtained by Payne and associates (1968) utilizing strain gages in their platform design.

Several limitations are present in the utilization of the force platform in biomechanics research. It is difficult to properly secure the platform to provide adequate damping. Because of the complexities of construction and instrumentation, most platforms to date have necessarily been limited in size. The dimensions are restricted further by the need for an acceptable frequency response level. For example, Payne and co-workers (1968) noted this problem by stating that the platform frequency response of 18 cycles/sec. was of the same order as the change in forces being measured. Since the units are relatively small, the sports movements which can be studied are restricted. Further, this technique is limited by the fact that the position and motion of the subject must be known before a thorough interpretation of the force recordings is possible.

In spite of the inherent problems associated with this technique, a number of sports skills have been investigated. Payne and associates (1968) and Payne singly (1968) conducted extensive analyses of the vertical jump, sprint start, running, shot put and weight lifting. Ramey (1970, 1972) examined the force characteristics of the take-off in the long jump and described how force plates can be utilized to investigate this sport skill. Carlsöö (1967, 1968), combining cinematography and electromyography, studied the ground reaction forces during swings of a highly skilled golfer. This combined approach provided considerably more information than use of any one system alone and will no doubt become more widely employed in the future. Force-time characteristics of the high jump take-off were studied by Kuhlow (1971) to discover the variations in force which occur during this phase of the movement. Differences were noted between good and poor performers as well as between the flop and straddle jumping styles.

The force platform as a useful tool continues to be developed and refined but has definitely taken its place in biomechanics research. As improvements occur in technology, this method will receive increasingly greater use in the study of sport movements.

Acceleration, an essential element in human performance, is of special interest because it is directly related to force ($\Sigma \mathbf{F} = m\mathbf{a}$) in linear terms and to torque ($\Sigma \mathbf{T} = I_o \boldsymbol{\alpha}$) in angular motion. Transducers developed to measure acceleration utilize this direct relationship with force. The general design of these devices, called accelerometers, consists of a mass suspended in a spring-damper system. The displacement of the mass is sensed by some means and an output proportional to the applied acceleration is produced. A variety of detectors are utilized in accelerometers but the most common are the variable reluctance sensor, the differential transformer, the piezoelectric device and the strain gage (Alnutt and Weinberg, 1963). Figure 7-4 depicts a strain gage type of accelerometer which was used as part of an elbow flexion test apparatus.

The major problem associated with acceleration measurement is caused by the normal gravitational influence. Alteration of the angle between the axis of the accelerometer and the horizontal causes a change in the component of the acceleration due to gravity. This difficulty can be partially overcome by use of a triaxial accelerometer which permits detection in three planes (Karger, 1958). A second approach involves the use of two accelerometers mounted in opposition so that the effect of gravity is cancelled (Figure 7-9). This principle was applied in the quick release method for estimating moment of inertia described in Chapter 5.

Since human limb motions are primarily angular in nature it is often desirable to measure angular acceleration (α). This can be done by using a linear accelerometer to detect the tangential acceleration (a_t) at a known distance (r) from the axis of rotation and converting this to angular units utilizing the relationship, $\alpha = a_t/r$. Direct measurement is also possible with an angular accelerometer (Norton, 1969).

The investigation of acceleration has received only limited attention from sport biomechanists. This is perhaps due to the difficulties encountered when movements are not in a horizontal plane and the limited number of limbs which can be monitored at one time. It is also necessary to locate precisely the positions of the body and its segments relative to the acceleration in order to properly interpret the results. The most effective use of acceleration measurement has been in the study of the mechanical characteristics of fundamental movements (see apparatus in Figure 7-4). Acceleration during a horizontal adductive movement was measured by Wartenweiler and Wettstein (1971) in their study of the characteristics of simple movements. In a previous study, Wartenweiler and co-workers (1970) examined the acceleration occurring in the lumbar region and at the hand during a shot put. Measurement of the acceleration of the boat during rowing was accomplished by Ishiko (1968) using telemetric techniques.

It is apparent that a variety of transducers are available for sensing biomechanical components of human movement. The devices most commonly used at present have been described but improved transducers are continually being developed and need to be evaluated. Norton has prepared an excellent list of the criteria for selection of transducers (1969, pp. 56–58). He discusses such factors as measurement requirements, system capability, transducer design and availability in detail.

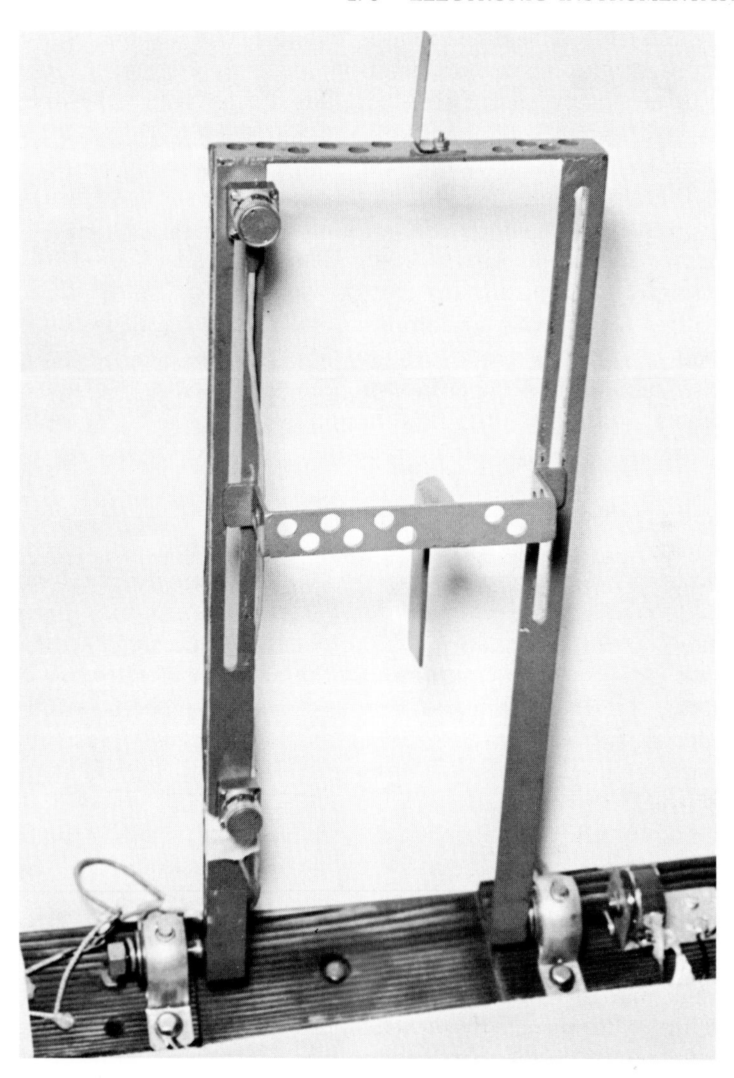

FIGURE 7-9. *Two-Accelerometer System Used to Cancel Effect of Gravity.*

Signal Conditioners. The electrical signal from the transducer is normally in a form unsuitable for display or recording. The signal is frequently very weak and often contains artifacts or extraneous information. The process of altering the signal is called signal conditioning (Honeywell Instrumentation Handbook, 1970) or modifying (Alnutt and Weinberg, 1963). Conditioners most commonly required in biomechanics research are amplifiers, integrators, differentiators, rectifiers and special signal processing components. These devices are located between the transducer and recording device in most instrumentation systems.

The amplifier is perhaps the most essential conditioner and serves to increase the voltage, current or power (capability of performing work) of an electrical signal. Amplifiers are normally classified on the basis of frequency response and the degree of input/output isolation. A DC amplifier is one providing a constant DC output level for constant DC input signal. Until recently the cost of a DC amplifier was prohibitive for general use but now inexpensive models of good quality are available. All other amplifiers are classified as AC and are normally reserved for applications such as ECG, EMG, EEG and AC-excited LVDTs and accelerometers in which DC response is not desired.

Amplifiers are also classified on the basis of their terminal arrangement and input/output isolation. A single-ended amplifier is one in which a common connection exists between one input and one output terminal. A differential amplifier has two isolated input terminals with one output terminal responsive to the signal difference between the input terminals. In practice amplifiers often represent a compromise between single-ended and true differential concepts.

Selection of the appropriate amplifier for a specific measurement problem entails an understanding of the principal features of those available. A DC amplifier is in many instances the best choice, provided use of neither EMG recording, LVDTs nor AC accelerometers is anticipated. There are three advantages of the DC amplifier (Honeywell Instrumentation Handbook, 1970). It provides for calibration of a static system, excellent low-frequency response and fast overload recovery. The latter feature is of major importance when high accuracy at low frequencies is required.

The differential amplifier, because of its input-to-output isolation, provides unmatched versatility. Hence, it is most often selected for multi-purpose use in biomechanics research. However, the single-ended amplifier, though less flexible, is often of lower cost and more reliable and, in certain situations, represents the best alternative. Evaluation of the many characteristics of amplifiers is necessary before a choice can be made. Since a detailed discussion of these characteristics is not feasible a condensed checklist is provided in Table 7-1.

Commonly used signal modifiers, in addition to amplifiers, include attenuators, differentiators, integrators, filter networks and analog-to-digital converters. An attenuator performs a function opposite that of an amplifier, namely, to decrease signal amplitude or power. Attenuators are normally inserted either at the input stage or between stages which control the gain of the amplifier.

Differentiators and integrators are used to modify signals as a

Table 7-1. Amplifier Specifications

Specifications	*Comments*
Type	AC or DC? Single-ended or differential?
Frequency Response	DC (low frequency) or AC? Controllable?
Linearity	Within acceptable limits over range of test values?
Gain	Magnitude and stability adequate for application?
Offset and Drift	Degree to which DC amplifier indicates zero with no signal (offset) or how this changes with time or temperature (drift)?
Input Impedance	Sufficiently high?
Output	Current and/or voltage capability sufficient for desired use? Magnetic tape recording capability?
Noise	Sufficiently low relative to signal being amplified?
Common Mode Rejection (CMR)	Ability of differential amplifier to ignore common mode signals?
Power Supply	Self-contained or several amplifiers operate off one supply?
Operational Environment	Influence of changes in temperature and relative humidity?
Size and Portability	Rack mounted or stand alone? Easily moved?
Cost	

preliminary step in data processing and analysis. For example, output from a potentiometer in the form of angular position/unit of time can be converted to angular velocity by differentiation. Likewise, acceleration with respect to time can be changed to velocity by integrating the signal from an accelerometer.

The use of the digital computer for data processing and analysis requires that analog data first be converted to digital form. This is performed by an analog-to-digital (A to D) converter, the output from which can be recorded on punched tape or transmitted directly to a digital computer.

Signal modifiers, then, are essential components to electronic measurement systems used in biomechanics research. They provide for changes in the transducer signal which are necessary before display and/or recording can take place. The specific modification depends upon the nature of the signal being processed and the desired outcome at the recording stage. Careful selection of the appropriate modifier system must be made if accurate results are to be secured.

Recorders and Display Devices. The last component of the measurement system is the recorder or display medium. Because a great variety of recording devices is available, the researcher may encounter difficulties in deciding upon the appropriate recorder to meet his specific needs. Consideration must be given to the detailed specifications, the compatibility with the transducer and modifier systems, the purchase price and the operating costs and maintenance of the recorder. The most commonly utilized recorder in biomechanics research is the graphic type, which includes recorders ranging from the slow-responding pen and ink devices to the high-speed light beam recorders. Those used to a lesser degree are the magnetic tape recorder and the cathode-ray oscilloscope.

The fundamental operating principles are demonstrated in a typical graphic recorder such as the direct writing galvanometric type. The components consist of a movable galvanometer, writing arm and stylus attached to the meter arm, chart paper and drive mechanism. The galvanometer responds directly to the changing voltage of the input signal producing a deflection in the writing arm. This deflection is recorded by the stylus on the moving chart paper. By altering the speed of the paper, the record can be expanded or condensed. Such a system provides for a permanent record of the phenomenon under investigation.

Perhaps the single most important feature of the recorder to consider is the frequency response. In general, the ink writer type of galvanometric recorders does not exceed about 150 Hz. Factors which affect the response time are the resonant frequency of the instrument and its mechanical characteristics including stylus friction, weight of the moving components, etc. A higher-frequency response is possible with the light beam oscillograph due to the lower inertia of the deflection elements. The light beam oscillograph also provides for overlapping of traces, which increases the versatility of this type of recorder.

Magnetic tape recording, although relatively expensive, provides the biomechanist with a very efficient means of recording and storing experimental data. One of the principal advantages of such a unit is that output from a number of measurement devices can be recorded simultaneously in real time with a minimum of instrumentation. Such a recorder performs the function of the data transmission or telemetry link and permits re-examination of the experimental data. Visual as well as audio display can be repeated as can computer manipulations of the recorded information. Utilizing different speeds for recording and reproducing, the investigator can alter the time base so that events occurring at a very rapid rate can be slowed down and examined more carefully.

The three most common magnetic tape recording techniques are as follows: direct (AM), indirect (FM) and digital. Required components for playback and recording include input and output signal conditioners, magnetic heads, tape and the tape transport system (Alnutt and Weinberg, 1963). The direct technique, which is the simplest but least accurate of the three methods, responds to changes in amplitude of the input signal. Two principal limitations of this technique are that slowly changing signals cannot be recorded and amplitude variations are subject to distortion in the recording process.

The indirect technique, referred to as FM, utilizes a carrier signal which is frequency modulated to respond to the input signal. This technique makes possible the recording of low-frequency input signals and is considerably more accurate than the direct recording method, even though high frequencies are difficult to record. The digital type has been less widely used, perhaps because of the problems of compatibility with instruments providing the input data. The digital format does eliminate the need for A to D conversion before computer data analysis can be performed.

Magnetic tape recording provides the capability of rapidly recording large quantities of data which can be replayed at slower speeds and edited for computer analysis. Its major contribution could be in research involving simultaneous measurement of a number of parameters such as angular displacement, acceleration and EMG. This technique has not been widely used by sports biomechanists to date, but will play an increasingly important role in the future.

Because of the vast number of recorder models available and the lack of standard terminology to describe them, it is necessary to study their principal features carefully so that direct comparisons can be made. Selection of the appropriate device for a particular situation can often be difficult. The specifications outlined in Table 7-2 should be considered in light of the planned use of the recorder. Technical assistance is imperative unless the researcher is well grounded in electronics.

MULTIPLE INSTRUMENT SYSTEMS

Recent advances in instrumentation have centered upon methods of securing larger quantities of data from a greater number of bio-

Table 7-2. Specifications for Recorders

Specifications	*Comments*
Type	Ink writing, light beam, magnetic tape, oscilloscope plus camera
Frequency Response (band width)	Adequate for application?
Slewing Speed	Time for full-range deflection related directly to frequency response?
Recording Medium	Paper, photo sensitive or wax paper, magnetic tape? Cost? Will it fade? Photographic developing required?
Number of Channels	Sufficient? Cost of additional channels?
Input Impedance	High enough?
Time Base	Adequate? Variable?
Accuracy and Repeatability	Acceptable for measurement?
Sensitivity	Fixed or variable?
Dynamic Range	Ratio of maximum signal to minimum signal detected.
Overshoot	
Width of Trace	

mechanical components which can be recorded and analyzed in a relatively short period of time. The desire to gain a more complete quantitative analysis of sports movements has led researchers to employ more than one measurement system. A number of combinations have been reported in the literature. These have involved force, acceleration, EMG, electrogoniometry and cinematography. The need for recording and analyzing larger quantities of data has stimulated development of on-line computer systems.

Researchers have supplemented electrogoniometry with simultaneous cinematographic recording to gain a more complete understanding of the movements being studied. Results from force-platform recordings become more useful when they are integrated with data secured from motion films (Carlsöö, 1967, 1968). Other combinations include acceleration and force measurements in rowing (Ishiko, 1971), force and cinematography in swimming (Goldfuss and Nelson, 1971) and force and electromyography in a knee extension test (Soderberg, 1971). These examples are representative of the many possibilities which exist for multiple-system recording.

One of the major deterrents to the development of biomechanics research has been the problem of processing and analyzing large quantities of data. Rapid strides have now been made through on-line computer systems such as those described by Petak (1971) and Sukop and his associates (1971). Figure 7-10 displays the components of the system reported by Sukop, Petak and Nelson to measure the force-time characteristics of isometric contractions. Force, detected by the LVDT unit, was sampled 30 times/sec., converted from analog to digital form and relayed to the computer. The computation of the parameters was accomplished in one second and the results were printed out immediately. A similar system was used by Soderberg (1971) to record knee extension force and electromyographic signals simultaneously during induced myotatic reflex (Figure 7-11). Recent studies by Lamoreux (1971) and Zuniga and associates (1972) have utilized on-line techniques in their investigations of human gait. The concept of immediate recording and analysis of biomechanical data will necessarily be expanded and refined because of the great savings in time and effort which are realized.

SUMMARY

Electronic instrumentation has become essential to the research capabilities of sport biomechanists. The authors have attempted to describe the essential characteristics of instrumentation systems

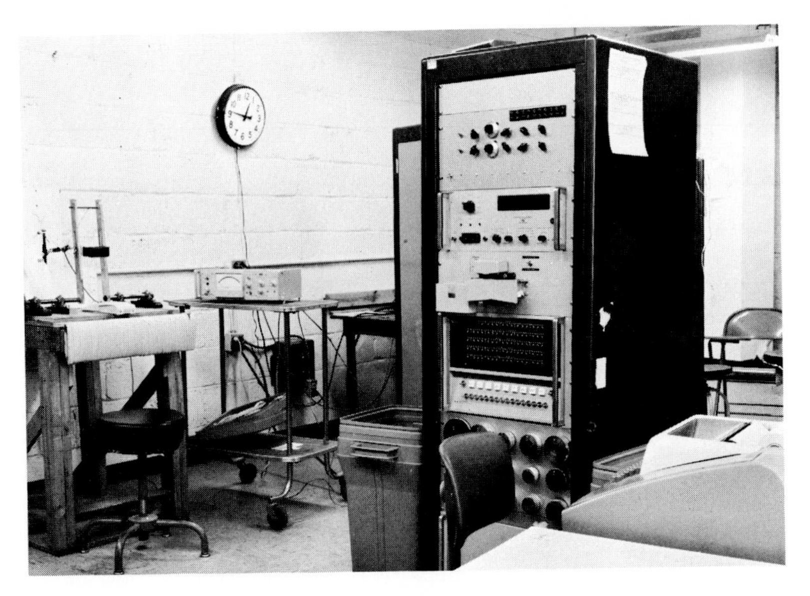

FIGURE 7-10. *On-line System for Recording and Analyzing Force-time Components of Isometric Contractions.*

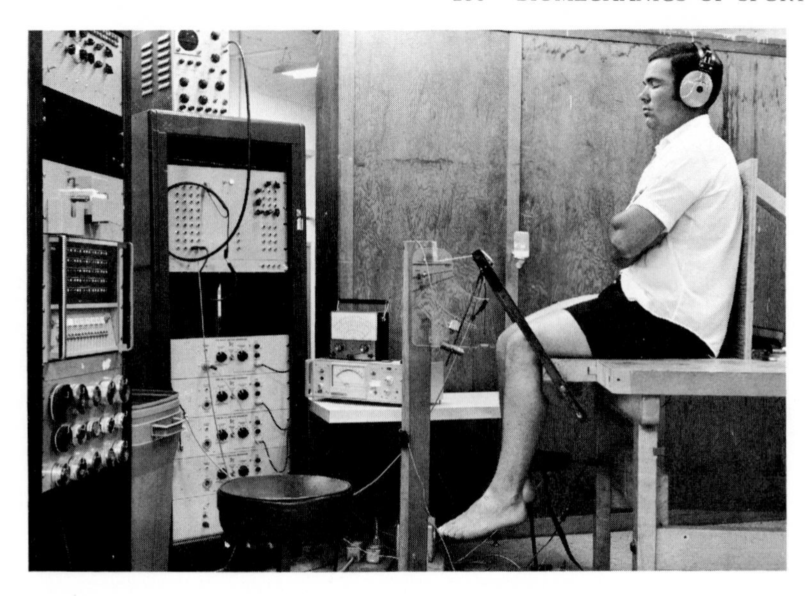

FIGURE 7-11. *System for Simultaneous Recording of Force and EMG Signals (From Soderberg, G. L.: Investigation of Selected Electrical Force and Time Characteristics of the Human Myotatic Reflex. Unpublished Doctoral Dissertation, Pennsylvania State University, 1971).*

including examples of application of their use. The material presented should be useful for persons actively involved in teaching and research concerning biomechanics.

SELECTED REFERENCES

Adamson, G. T., and Whitney, R. J.: Critical Appraisal of Jumping as a Measure of Human Power. In J. Vredendregt and J. Wartenweiler (Eds.), *Biomechanics II*. Basel: Karger, 1971.

Adrian, M. J.: An Introduction to Electrogoniometry. In *Kinesology Review 1968*. Washington: AAHPER, 1968.

Adrian, M. J.: Application of Electrogoniometry, Proc. Symposium Biomed. Engineering, Milwaukee, 98–101, 1966.

Alnutt, R., and Becker, W. C.: Techniques of Physiological Monitoring. Volume III Systems, RCA Service Co., October, 1964 (AD 609 481).

Alnutt, R., and Weinberg, P. T.: Techniques of Physiological Monitoring. Volume II Components, RCA Service Co., November, 1963 (AD 426 816).

Alt, F. (Ed.): *Advances in Bioengineering and Instrumentation*. New York: Plenum Press, 1966.

Barlow, D. A., and Cooper, J. M.: Mechanical Considerations in the Sprint Start. Athletic Asia, *2*, 27–35, 1972 (August).

Baumann, W.: Uber Ortsfeste und Telemetrische Verfahren zur Messung der

Abstosskraft des Fusses. In J. Wartenweiler, *et al.* (Eds.), *Biomechanics.* Basel: Karger, 1968.

Blader, F. B.: The Analysis of Movements and Forces in the Sprint Start. In J. Wartenweiler, *et al.* (Eds.), *Biomechanics.* Basel: Karger, 1968.

Blader, F. B., and Payne, A. H.: Instrumented Starting Blocks. Athletic Coach, *2,* 3–4, 1967.

Brendel, A. E.: The Structures of Strain Gauge Transducers: An Introduction. IEEE Transactions on Industry and General Applications. *IGA-5,* pp. 90–94, 1969 (January–February).

Brooks, C. M.: Validation of the Gamma Mass Scanner for Determination of Center of Gravity and Moment of Inertia of Biological Tissue. Unpublished Master's Thesis, Pennsylvania State University, 1972.

Brophy, J. J.: *Basic Electronics for Scientists.* New York: McGraw-Hill, 1966.

Carlsöö, S.: Kinematic Analysis of the Golf Swing. In J. Wartenweiler, *et al.* (Eds.), *Biomechanics.* Basel: Karger, 1968.

Carlsöö, S.: A Kinetic Analysis of the Golf Swing. J. Sport Med. *7,* 76–82, 1967 (June).

Carlsöö, S.: A Method for Studying Walking on Different Surfaces. Ergonomics, *5,* 271–274, 1962.

Goldfuss, A. J., and Nelson, R. C.: A Temporal and Force Analysis of the Crawl Arm Stroke during Tethered Swimming. In L. Lewillie and J. P. Clarys (Eds.), *Biomechanics in Swimming.* Brussels: Universite Libre de Bruxelles, 1971.

Gollnick, P. D., and Karpovich, P. V.: Electrogoniometric Study of Locomotion and of Some Athletic Movements. Res. Q. Amer. Assoc. Health Phys. Ed., *35,* 357–369, 1964.

Gombac, R.: The Mechanics of Take-off in High Jump. In J. Vredenbregt and J. Wartenweiler (Eds.), *Biomechanics II.* Basel: Karger, 1971.

Greene, J. H., and Morris, W. H.: The Design of a Force Platform for Work Measurement. J. Industr. Engin., *10,* 312–317, 1959 (July–August).

Greene, J. H., and Morris, W. H.: The Force Platform—An Industrial Engineering Tool. J. Industr. Engin., *9,* 131–132, 1958 (March–April).

Grieve, D. W.: A Device called the Polgon for the Measurement of the Orientation of Parts of the Body Relative to a Fixed External Axis. J. Physiol. *201,* 70, 1969.

Hearn, N. K. H., and Konz, S.: An Improved Design for a Force Platform. Ergonomics, *11,* 383–389, 1968.

Henry, F. M.: Factorial Structure of Speed and Static Strength in a Lateral Arm Movement. Res. Q. Amer. Assoc. Health Phys. Ed., *31,* 440–447, 1960.

Henry, F. M.: Force-time Characteristics of the Sprint Start. Res. Q. Amer. Assoc. Health Phys. Ed., *23,* 301–310, 1952.

Henry, F. M., and Whitley, J. D.: Relationships Between Individual Differences in Strength, Speed, and Mass in an Arm Movement. Res. Q. Amer. Assoc. Health Phys. Ed., *31,* 24–33, 1960.

Honeywell Instrumentation Handbook, Denver: Honeywell, Inc., Test Instruments Division, 1970.

Hutinger, P. W.: Construction and Utilization of a Simple Electrogoniometer. In J. M. Cooper (Ed.), *Selected Topics on Biomechanics.* Chicago: Athletic Institute, 1971.

Ishiko, T.: Biomechanics of Rowing. In J. Vredenbregt and J. Wartenweiler (Eds.), *Biomechanics II*. Basel: Karger, 1971.

Ishiko, T.: Application of Telemetry to Sports Activities. In J. Wartenweiler, *et al.* (Eds.), *Biomechanics*. Basel: Karger, 1968.

Johnson, R. C., and Smidt, G. L.: Measurement of Hip-Joint Motion During Walking. J. Bone Joint Surg., *51*A, 1083–1094, 1969.

Karger, D. W.: Instrumentation for Automatic Data Processing in Motion-Time Research. J. Industr. Engin., *9*, 123–127, 1958 (March–April).

Karpovich, P. V.: Electrogoniometry: A New Device for Study of Joints in Action. Fed. Proc., *18*, 79, 1959.

Karpovich, P. V., Herden, E. L., and Asa, M. M.: Electrogoniometric Study of Joints. U.S. Armed Forces Med. J., *11*, 424–450, 1960.

Klissouras, V., and Karpovich, P. V.: Electrogoniometric Study of Jumping Events. Res. Q. Amer. Assoc. Health Phys. Ed., *38*, 41–47, 1967.

Koniar, M.: The Electrogoniographic Method and its Position in the Methodology of Biomechanical Research on Human Motion. In J. Wartenweiler, *et al.* (Eds.), *Biomechanics*. Basel: Karger, 1968.

Korb, R. J.: A Simple Electrogoniometer: A Technical Note. Res. Q. Amer. Assoc. Health Phys. Ed., *41*, 203–204, 1970.

Korn, J., and Simpson, K.: Theory and Experiments for Teaching Measurement. Instrument Practice, Part 2, 756–763, 1970.

Korn, J., and Simpson, K.: Theory and Experiments for Teaching Measurement. Instrument Practice, Part 4, 43–46, 1971a.

Korn, J., and Simpson, K.: Theory and Experiments for Teaching Measurement. Instrument Practice, Part 6, 163–168, 1971b.

Kuhlow, A.: *Analyse moderner Hochsprungtechniken*. Sportwissenschaftliche Arbeiten Band 5. Berlin: Bartels & Wernitz KG, 1971.

Lamoreux, L. W.: Kinematic Measurements in the Study of Human Walking. Bull. Prosthet. Res., 10–15: 3–84, 1971 (Spring).

Magel, J. R.: Propelling Force Measured during Tethered Swimming in the Four Competitive Swimming Styles. Res. Q. Amer. Assoc. Health Phys. Ed., *41*, 68–74, 1970.

Maier, I: Measurement Apparatus and Analysis Methods of the Biomotor Process of Sport Movements. In J. Wartenweiler, *et al.* (Eds.), *Biomechanics*. Basel: Karger, 1968.

Mitchelson, D. L.: An Opto-electric Technique for Analysis of Angular Movements. Master of Science Thesis, Loughborough University of Technology, England, 1971.

Murray, M. P., Seireg, A., and Scholz, R. C.: Center of Gravity, Center of Pressure and Supportive Forces during Human Activities. J. Appl. Physiol., *23*, 831–838, 1967.

Nelson, R. C., and Fahrney, R. A.: Relationship Between Strength and Speed of Elbow Flexion. Res. Q. Amer. Assoc. Health Phys. Ed., *36*, 455–463, 1965.

Nelson, R. C., Larson, G., Crawford, C., and Brose, D.: Development of a Ball Velocity Measuring Device. Res. Q. Amer. Assoc. Health Phys. Ed., *37*, 150–155, 1966.

Norton, H. N.: *Handbook of Transducers for Electronic Measuring Systems*. Englewood Cliffs, N.J.: Prentice-Hall, 1969.

O'Leary, J. P.: A Strain-gauge Force Platform for Studying Human Movement. Percept. Motor Skills, *30*, 698, 1970.

Payne, A. H.: The Use of Force Platforms for the Study of Physical Activity. In J. Wartenweiler, *et al.* (Eds.), *Biomechanics*. Basel: Karger, 1968.

Payne, A. H. and Blader, F. B.: The Mechanics of the Sprint Start. In J. Vredenbregt and J. Wartenweiler (Eds.), *Biomechanics II*. Basel: Karger, 1971.

Payne, A. H., Slater, W. J., and Telford, T.: The Use of a Force Platform in the Study of Athletic Activities. A Preliminary Investigation. Ergonomics, *11*, 123–143, 1968.

Petak, K. L.: The Acquisition and Reduction of Biomechanical Data by Minicomputer. In J. M. Cooper (Ed.), *Selected Topics on Biomechanics*. Chicago: Athletic Institute, 1971.

Ramey, M. R.: Effective Use of Force Plates for Long Jump Studies. Res. Q. Amer. Assoc. Health Phys. Ed., *43*, 247–252, 1972.

Ramey, M. R.: Force Relationships of the Running Long Jump. Med. Sci. Sports, *2*, 146–151, 1970.

Reed, D. J., and Reynolds, P. J.: A Joint Angle Detector. J. Appl. Physiol. *27*, 745–748, 1969.

Ringer, L. B., and Adrian, M. J.: An Electrogoniometric Study of the Wrist and Elbow in the Crawl Arm Stroke. Res. Q. Amer. Assoc. Health Phys. Ed., *40*, 353–363, 1969.

Scheuchenzuber, H. J.: A Biomechanical Analysis of Four Backstroke Starts. Unpublished Master's Thesis, Pennsylvania State University, 1970.

Soderberg, G. L.: Investigation of Selected Electrical, Force and Time Characteristics of the Human Myotatic Reflex. Unpublished Doctoral Dissertation, Pennsylvania State University, 1971.

Stothart, J. P.: A Biomechanical Analysis of Static and Dynamic Muscular Contraction. Unpublished Doctoral Dissertation, Pennsylvania State University, 1970.

Sukop, J., Petak, K. L., and Nelson, R. C.: An On-line Computer System for Recording Biomechanical Data. Res. Q. Amer. Assoc. Health Phys. Ed., *42*, 101–102, 1971.

Taylor, B. M., and Karpovich, P. V.: Electrogoniometry—A Study of Man in Motion. Canad. Assoc. Health Phys. Ed. Rec., *32*, 17–18, 1965 (October–November).

Wartenweiler, J., Wettstein, A., and Lehmann, G.: Shot Put: Subjective and Objective Aspects of Motor Design. In G. S. Kenyon (Ed.), *Contemporary Psychology of Sport*. Chicago: Athletic Institute, 1970.

Wartenweiler, J. and Wettstein, A.: Basic Kinetic Rules for Simple Human Movements. In J. Vredenbregt and J. Wartenweiler (Eds.), *Biomechanics II*. Basel: Karger, 1971.

Whitley, J. D. and Smith, L. E.: Velocity Curves and Static Strength-action Strength Correlations in Relation to the Mass Moved by the Arm. Res. Q. Amer. Assoc. Health Phys. Ed., *34*, 379–395, 1963.

Zuniga, E. N., Leavitt, L. A., Calvert, J. C., Canzoneria, J., and Peterson, C.: Gait Patterns in Above Knee Amputees. Arch. Phys. Med., *53*, 373–382, 1972.

CHAPTER 8

Digital Computer Techniques

ALTHOUGH of comparatively recent origin, the high-speed digital computer has made an immeasurable impact upon society. It has become indispensable in engineering, business and industry and is making increasing contributions in the fields of medicine and education. The services of the digital computer have been employed at all levels of research beginning with an investigation of the literature, through data collection, reduction and statistical analysis to the graphical display of results. Because of the importance and widespread use of the computer, familiarity with the capabilities of the various types of computers and an understanding of programming principles have become a necessity for the specialist in biomechanics.

HARDWARE SYSTEMS

The computer market continues to undergo extensive expansion. In response to the demand, several different hardware systems have evolved to serve specific needs. Analog computers process data of a continuous nature such as voltage signals while digital computers deal with discrete bits of information. A hybrid machine, as the name implies, includes both analog and digital features. The digital type has been utilized most often in biomechanics research since it is an extremely efficient calculating device capable of fundamental arithmetic and logic operations which, in combination, can provide the solution to extremely complex problems. High-speed electronic digital computers are manufactured in various sizes, shapes and price ranges. They include the large models commonly installed on a monthly rental basis at universities and in industry; minicomputers utilized in on-line applications in laboratories (Sukop *et al.*, 1971; Petak, 1971); and most recently, the small, inexpensive programmable calculators.

The advent of remote terminals and time sharing in large computer systems, and the small digital computers have led to unique possibilities for measuring and recording biomechanical data. The relative merits of these two are of special importance to persons developing a laboratory program because a system in which biomechanical parameters can be immediately measured, recorded and prepared for statistical analysis offers a new dimension in biomechanics research.

The remote terminal in communication with a large computer having time-sharing capability is one possibility for on-line data recording. However, there are a number of limitations which make it ineffective as an on-line system operating in real time. Because the interaction from the terminal to the main computer is on a time-sharing basis, attention required for the conduct of a specific experiment cannot be assured. Although the normal telephone lines can handle data from a teletype or tape reader, the higher transmission rates often encountered in biomechanics experiments require special equipment available only at a much higher rental rate. Therefore, the possibility that the computer would act as a controller by interacting with a subject or experimental equipment is virtually impossible from a remote terminal. However, this system does offer some special capabilities provided that real-time, on-line interaction is not required. Availability of large computers with an almost infinite memory capacity and accessibility of library subroutines for most mathematical and statistical computations permit rather extensive and complex programs to be used. Input from a teletype, card or paper tape reader offers considerable flexibility to the user. Normal telephone lines provide easy connection to the main computer center. In most cases, though, these represent convenience and, to some extent, merely increased speed of data processing. Most of these functions can be performed by taking the data cards or tapes to the computation center for processing. Whether the added cost of the remote terminal can be justified on the basis of convenience may be questionable.

The most promising system to date for real-time data recording has been the small, stand-alone digital computer sometimes referred to as the mini-computer. Its principal advantage lies in the processing of data at high transmission rates in real time. Also, the computer has the capacity to perform a controlling function by interacting directly with the subject or experimental equipment. It has been used successfully for this purpose and meets all the present needs for real-time, on-line recording. Future developments will no doubt lead to the full realization of the vast potential of the on-line concept in biomechanics. Secondary benefits include the use of the mini-computer for teaching graduate students. The possibility of students actually operating the computer offers a considerably more interesting learning environment than that generally associated with a large computation center. The small digital computer can also function as a programmable calculator when it is not being used in an on-line mode. Although the memory of the typical unit is somewhat limited (between 4K and 8K), it can be expanded at a reasonable cost. The initial price of the mini-computer is relatively high ($10,000 to $15,000) but its life span is generally long. In comparison, the cost is roughly equivalent to rental of a remote terminal for three or four years. In order to develop fully the potential of the system, mechanical, electronic and computer programming expertise must be available.

Rapid advances and improvements in electronic programmable calculators also have important implications for biomechanics research. As the name suggests, the main difference between these calculators and the manual type is that they have a memory unit which permits them to store and execute a program or series of instructions. In function, they resemble the larger digital computers. Because their storage capacity for instructions and data is relatively limited, they are restricted in the length and the complexity of the programs which they can handle. Nevertheless, they are quite capable of performing many of the mathematical operations commonly used in biomechanics research. These include center of gravity determinations, conversion of film distances and velocities to real-life dimensions, prediction of segmental parameters from regression equations and many types of statistical analyses.

Most programmable calculators can be expanded by the addition of optional peripheral devices such as printers, card readers, x-y plotters, extra memory blocks and units to convert analog signals into digital form. Thus, a fairly comprehensive computing system which reflects the needs and interests of the user can be built up over the course of a few years as funds become available from budget and grant sources. When purchasing one of these machines, the user should consider factors such as ease and simplicity of programming, expandability, memory size, and the availability of service personnel.

DATA REDUCTION AND PROCESSING

Computer data processing is an integral part of research in biomechanics. The large volume of numerical information and extensive mathematical operations which often accompany biomechanical investigations make it a virtual necessity. Most raw data require some type of conversion to put them into a meaningful form for interpretation. These transformations and subsequent calculations can be performed quickly and accurately by utilizing a high-speed digital computer. Many higher-level programming languages which serve as a translation link between man and machine have been developed for this purpose. One of the most common, called FORTRAN, mnemonic for formula translation, closely resembles mathematical notation and permits the user to communicate with the computer with relative ease. When writing the necessary sets of computer instructions, referred to as computer programs, the user must learn and carefully observe certain rules of syntax specific to the particular language. Regardless of the language chosen, however, common procedures apply to the planning, documentation and verification of the programs which are developed.

A little effort devoted to planning the most efficient method of recording the experimental data and designing a program to yield maximum information from the input supplied will be amply repaid. It is advisable to experiment with a small amount of data before making a final decision on the best format. An attempt should be made to record all the important information while, at the same time, avoiding unnecessary duplication of effort. For example, the coordinates of segmental endpoints taken from film can be used to calculate the center of gravity location, limb positions and joint angles. The order for recording these points should minimize extraneous movements of the analyzer controls and should be maintained if the raw data have to be transferred onto cards or paper tape. Adequate space should be left between the numbers to make checking easier. If punched cards are employed as the computer input medium, key punch machines should be programmed to leave appropriate blanks, position the numbers in the correct locations and automatically duplicate pertinent identification on each card.

While the time required to provide careful and complete documentation may appear unnecessary when the program is first written, its value will be appreciated when the program is run at some future date. Toward this end, it is strongly recommended that variables be labelled mnemonically to facilitate their identification in the program. Thus, the component of velocity along the X axis might be referred to as VELX, XVEL, VX or XV rather than an arbitrary notation such as B or C. A list of these variables, their designations in the program and their units should also be kept. Generous use of comments throughout the program will help to record the logic of the calculations. In the case of card input, the numbering of the cards in sequence is a highly recommended practice as it ensures that the proper order of statements is maintained. Since corrections and additions will undoubtedly be made, the date of the current revision should be included in the program listing. It is also advisable to have at least one duplicate copy of the computer program and the essential data as insurance against loss or damage of the original.

When processing data by computer, the investigator must take care to achieve accurate results. Errors in transcription can be reduced by comparing the data input to the computer with the original figures. Since there is usually a logical sequence to the values of a given variable, any which do not appear to fit this pattern should be double checked. Incorrect output may be caused by errors in logic in the program which are sometimes difficult to identify. In attempting to locate the source of the problem, it may be helpful to have the results printed out at intermediate points in the program. Initial calculations should be verified by hand. Once the program is running correctly, test data should be input and a record kept of the results. These data may then be utilized to check the program before it is used at a later date or after a duplicate copy has been made.

LIBRARY PROGRAMS

Although it is common practice to write a computer program to meet the requirements of a specific biomechanics project, library programs may also be used. The latter have been developed for most statistical applications and other routine mathematical operations including curve fitting and numerical analysis. Both the Scientific Subroutine Package (SSP) and the Statistical Package for the Social Sciences (SPSS) contain numerous library programs with detailed descriptions of the formulae employed, specific instructions for their use and possible limitations in application. These and other similar programs are available at local computation centers.

While statistical and mathematical operations can be carried out rapidly by computer, the user must understand the theory behind such calculations. Otherwise he will tend to accept answers generated by the computer without question when in fact they may be incorrect. Failure to follow the library program instructions precisely or errors in the input data values will lead to invalid results. Therefore, the user should hand calculate a sample of the data and become familiar with the range of answers expected. A careful inspection of the total computer output will also help to spot possible inaccuracies.

Since library programs are written for use by a large number of individuals, they are general in scope, account for extremes in data values, include many internal checks and are usually quite well validated and documented. A word of caution, however, should be mentioned about using computer programs developed by other investigators which do not meet these specifications. When a researcher writes his own program, he usually intends it for a particular application and probably for data values within certain ranges. He is familiar with its limitations and knows where to make minor alterations to change the output. His documentation, while adequate for his own purposes, may not be sufficient for someone unfamiliar with the program. In this instance, the potential user is better advised to write his own program than to spend the time and effort required to decipher someone else's routines, make them compatible with his own computer system, change the input and output formats to meet his needs and verify the calculations.

COMPUTER GRAPHICS

Computer graphics, a technique for displaying computer output pictorially which has won wide acclaim in the fields of architecture and design, has recently been used to illustrate the results of biomechanical investigations. Perhaps its greatest potential lies in the effective graphical presentation of the output from simulation studies of human motion (Figure 8-1). Computer-generated illustrations drawn on paper with an incremental plotting device provide a permanent record or "hard copy" which is useful for graphs and pictures of various biomechanical parameters. If several displays are to be presented in a limited space, as in the case of a human movement sequence involving little translation, the overlapping of images may present a problem. This can be overcome by using a cathode ray tube as the output medium. Since the light beam draws the picture very rapidly, there is an illusion of continuous motion. Some of these displays are interactive, and permit the operator to communicate directly with the computer and alter the image in an on-line fashion.

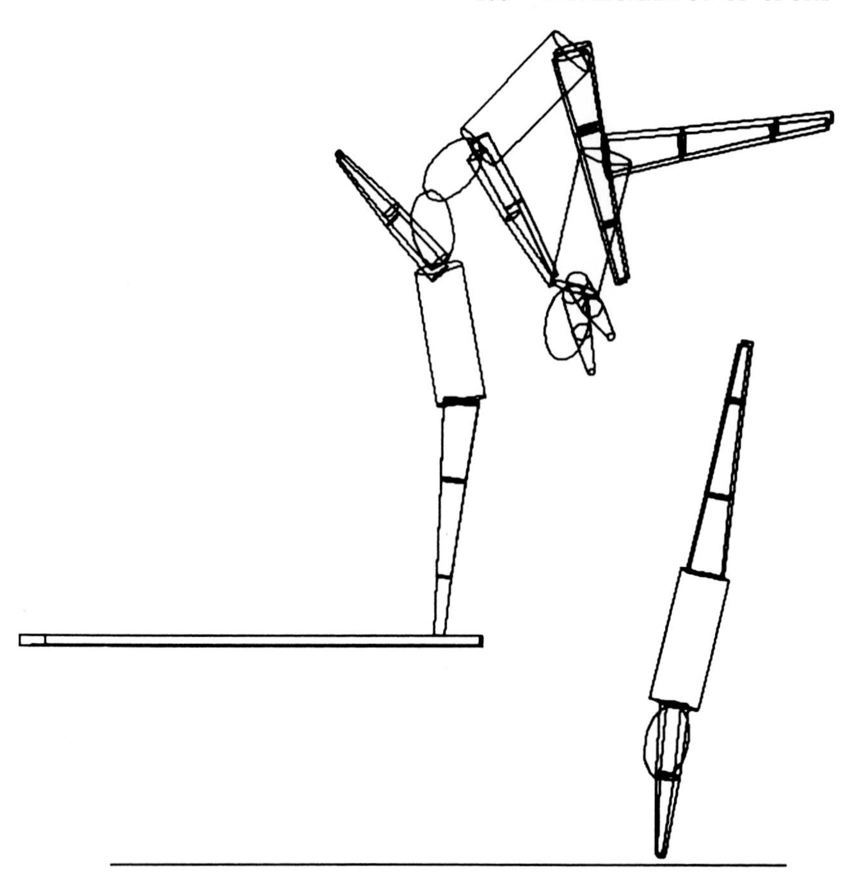

FIGURE 8-1. *Computer Graphics Display.*

To construct the desired graphical representations, the computer must be supplied with instructions in the form of a computer program which designates the position of the pen or light beam in a rectangular reference frame. Usually a set of coordinates is given in either two- or three-dimensional space. The pen or beam then moves to that location in a writing or nonwriting mode as specified by the program instructions. Most graphics systems are capable of generating these directed line segments or vectors. The inclusion of curves, which enhances the appearance of the illustration, requires a series of short straight lines. The shorter they are, the smoother the curve. From these basic building blocks, rather complex drawings of the human body and other biomechanical systems can be constructed (Fetter, 1967; Garrett *et al.*, 1968).

Software packages which provide the necessary routines for computer graphics are generally based upon similar principles. Some have extensions for drawing in the third dimension. Most permit the whole diagram and/or its components to be translated, rotated and scaled to appropriate sizes. While slight variations will be encountered in the different systems, the basic principle of drawing within an established coordinate system remains the same.

COMPUTER SIMULATION

One of the most recent applications of computer technology to the area of biomechanics has been that of simulation. It involves the development of a simplified mathematical model consisting of a series of appropriate equations which represent a biomechanical system. The system, composed of functionally related components, may be the total body of an athlete, the arms and club of a golfer, the swing leg of a sprinter or a discus in flight. The model is then translated into a suitable language so that it can be programmed for solution on a digital computer. After being validated against experimental data, it is used to simulate the behavior of the system under various carefully controlled conditions.

As with other methods of research, computer simulation has both advantages and limitations. It is particularly well suited for analysis involving extensive mathematical operations since the computer can perform the necessary calculations quickly and accurately. Equations which were virtually impossible to solve by other means because of the man-hours and even man-years required can now be done by computer within a matter of seconds. This capability permits a greater flexibility in model construction than would otherwise be possible. Because the model is free from the physical limitations commonly associated with experimental research, variables of interest can be isolated and carefully controlled. Their influence upon the performance of a motor skill can be investigated either singly or in combination with other factors.

Although this research technique promises to increase our knowledge of the mechanical basis of human motion, the construction of a valid simulation model to represent a particular biomechanical system adequately is a rather difficult task. Since human motion is far too complex to be perfectly duplicated on a digital computer, simulations must incorporate several simplifying assumptions. Thus, the body itself may be replaced by a series of rigid links whose biomechanical properties can be determined mathematically (Hanavan, 1964). The number of segments is usually reduced to a minimum and movement is often restricted to a single plane. The situation is obviously an artificial one. The objective in constructing a simulation model, however, is to achieve an optimum combination of accuracy and simplicity. This cannot be easily accomplished.

Simulations can be classified as either stochastic or deterministic. The former include probability functions and chance relationships (Lindsey, 1959, 1961, 1963). Inferential statistics plays an important role in their development. Therefore, they are appropriate for studying strategy in games like baseball, basketball, football and hockey for which elaborate statistics are kept. Biomechanical systems, on the other hand, are better investigated using the deterministic type of model which is based upon exact relationships. Thus, if $F = m * a$, and both mass and acceleration are known quantities, the value of the force can be calculated with certainty. The deterministic simulation tends to be less complicated than the stochastic and provides a unique result for a given set of input conditions.

Although the simulation method has been used to advantage in the engineering and business fields for some time, it has only been since the late 1960's that it has begun to take its place as a research technique for investigating the mechanical factors affecting performance in sport. Some studies have been concerned with the effect of external forces. Examples are Baumann's (1971) model, which described the mechanical factors influencing a toboggan run, and the one developed by Miller (1971) to depict the flight portion of nontwisting dives from a springboard or tower. Two segment computer models have been reported by Cochran and Stobbs (1968) and Jorgensen (1970) to investigate the dynamics of the golf swing. A simplified five-link system to analyze swimming mechanics has been proposed by Seireg and Baz (1971). Other investigators have considered the effect of internal muscle and joint forces in addition to the external forces. Chaffin's mathematical representation of lifting tasks illustrates this approach (Chaffin, 1969; Chaffin and Baker, 1970). The same basic model was used to estimate ankle torques, gastrocnemius tension and heel lift forces in downhill skiing positions (Quigley and Chaffin, 1971). Research on locomotion (Beckett and Chang, 1968; Chen and Huang, 1970) also falls into this category. Many of these models are yet in the initial stages and further modifications and improvements can be expected.

The ultimate contribution of the computer simulation method of research lies in the realm of prediction. How is the performance influenced by altering such factors as force, limb velocity, air resistance, friction, body proportions or take-off position? Can an optimum combination of movements be developed to accomplish a specific objective in sport? While the ultimate realization of this goal has not yet been achieved, computer simulation does provide increased understanding of the mechanics of human motion. The very exercise of attempting to construct a valid model forces a disciplined approach to the analysis. Information must be expressed in quantitative terms if it is to be included in the model. Gaps in knowledge concerning the activity become apparent during the process and new questions arise. Even though an ideal mathematical representation of the operation of the system may not be the immediate result of such research, important insights into the mechanics of man in motion are gained (Miller, 1973, In Press).

SELECTED REFERENCES

Barton, R. F.: *A Primer on Simulation and Gaming*. Englewood Cliffs, N.J.: Prentice-Hall, 1970.

Baumann, W.: The Influence of Mechanical Factors on Speed at Tobogganing. Paper presented at the Third International Seminar on Biomechanics, Rome, 1971.

Beckett, R., and Chang, K.: An Evaluation of the Kinematics of Gait by Minimum Energy. J. Biomechanics, *1*, 147–159, 1968.

Brick, D. B., and Chase, E. N.: Interactive CRT Display Terminals. Part 1—The Terminology & the Market. Mod. Data, *3*, 62–69, 1970 (May).

Brick, D. B., and Chase, E. N.: Interactive CRT Display Terminals. Part 3—Graphic CRT Terminals. Mod. Data, *3*, 60–68, 1970 (July).

Briggs, W. G.: *Baseball-o-mation, A Simulation Study*. Boston: Harbridge House, 1960.

Chaffin, D. B.: A Computerized Biomechanical Model—Development of and Use in Studying Body Actions. J. Biomech., *2*, 429–441, 1969.

Chaffin, D. B., and Baker, W. H.: A Biomechanical Model for Analysis of Symmetric Sagittal Plane Lifting. AIIE Trans. Indust. Eng. Res. Dev., *2*, 16–27, 1970.

Chase, E. N., and Brick, D. B.: Interactive CRT Display Terminals. Part 2—Alphanumeric CRT Terminals. Mod. Data, *3*, 70–85, 1970 (June).

Chen, Y. R., and Huang, B. K.: Kinematics and Computer Simulation of Human Walking. Paper No. 70-663, Am. Soc. Ag. Eng., 1970.

Cochran, A., and Stobbs, J.: *The Search for the Perfect Swing*. Philadelphia: Lippincott, 1968.

Cook, E.: *Percentage Baseball*. Cambridge, Mass.: MIT Press, 1966.

Davies, R. G.: *Computer Programming in Quantitative Biology*. New York: Academic Press, 1971.

Dutton, J. M., and Starbuck, W. H.: *Computer Simulation of Human Behavior*. New York: Wiley, 1971.

Eshkol, N. *et al.*: Notation of Movement. Final Report, Department of Electrical Engineering, University of Illinois, 1970 (AD 703 936).

Faiman, M., and Nievergelt, J. (Eds.): *Pertinent Concepts in Computer Graphics*. Urbana, Ill.: University of Illinois Press, 1969.

Fetter, W. A.: Computer Graphics. Design Quarterly, *66/67*, 14–23, 1967.

Fischer, B. O.: Analysis of Spinal Stresses During Lifting—A Biomechanical Model. Unpublished Master's Thesis, University of Michigan, 1967.

Fox, L., and Mayers, D. F.: *Computing Methods for Scientists and Engineers.* Oxford: Clarendon Press, 1968.

Garrett, G. E., and Reed, W. S.: Computer Graphics: Simulation Techniques and Energy Analysis. In J. M. Cooper (Ed.), *Selected Topics on Biomechanics.* Chicago: Athletic Institute, 1971.

Garrett, R. E., Widule, C. J., and Garrett, G. E.: Computer-Aided Analysis of Human Motion. In *Kinesiology 1968.* Washington: AAHPER, 1968.

Hanavan, E. P.: A Mathematical Model of the Human Body. Wright-Patterson Air Force Base, Ohio, 1964 (AMRL-64-102).

Hatze, H.: Optimization of Human Motions. Paper presented at the Third International Seminar on Biomechanics, Rome, 1971.

Hauser, S. M., Angel, R. K., and Holt, W. H.: Dedicated or Time Share? Res. Develop., *21,* 27–31, 1970 (April).

Jorgensen, T.: On the Dynamics of the Swing of a Golf Club. Amer. J. Phys., *38,* 644–651, 1970.

Kahne, S., and Salasin, J.: Computer Simulation in Athletic Performance. In the Proceedings of the Annual Meeting of the National College Physical Education Association—Men, Chicago, 1969.

Lindsey, G. R.: An Investigation of Strategies in Baseball. Oper. Res., *11,* 477–501, 1963.

Lindsey, G. R.: The Progress of the Score During a Baseball Game. Amer. Stat. Ass. J., *56,* 703–728, 1961.

Lindsey, G. R.: Statistical Data Useful for the Operation of a Baseball Team. Oper. Res., *7,* 197–207, 1959.

Miller, D. I.: Computer Simulation of Human Motion. In H. T. A. Whiting (Ed.), *Techniques for the Analysis of Human Movement.* London: Henry Kimpton, 1973. In Press.

Miller, D. I.: A Computer Simulation Model of the Airborne Phase of Diving. In J. M. Cooper (Ed.), *Selected Topics on Biomechanics.* Chicago: Athletic Institute, 1971.

Miller, D. I.: A Computer Simulation Model of the Airborne Phase of Diving. Unpublished Doctoral Dissertation, Pennsylvania State University, 1970.

Milsum, J. H., Kearney, R. E., and Kwee, H. H.: Signal Processing by On-Line Computer for Bio-mechanical Studies. Paper presented at the Third International Seminar on Biomechanics, Rome, 1971.

Occhiogrosso, J. J.: Interfacing—To Desk-Top Calculators. Res. Develop., *22,* 22–25, 1971 (April).

Pearson, J. R., McGinley, D. R., and Butzel, L. M.: A Dynamic Analysis of the Upper Extremity: Planar Motions. Hum. Factors, *5,* 59–70, 1963.

Peddie, J. G.: Interfacing—To a Remote Computer. Res. Develop., *22,* 26–28, 1971 (April).

Petak, K. L.: The Acquisition and Reduction of Biomechanical Data by Minicomputer. In J. M. Cooper (Ed.), *Selected Topics on Biomechanics.* Chicago: Athletic Institute, 1971.

Prince, M. D.: *Interactive Graphics for Computer-Aided Design.* Reading, Mass.: Addison-Wesley, 1971.

Quigley, B. M., and Chaffin, D. B.: A Computerized Biomechanical Model Applied to the Analysis of Skiing. Med. Sci. Sports, *3,* 89–96, 1971.

Reed, W. S., and Garrett, R. E.: A Three-Dimensional Human Form and Motion Simulation. In *Kinesiology 1971*. Washington: AAHPER, 1971.

Secrest, D., and Nievergelt, J. (Eds.): *Emerging Concepts in Computer Graphics*. New York: W. A. Benjamin, 1968.

Seireg, A., and Baz, A.: A Mathematical Model for Swimming Mechanics. In L. Lewillie and J. P. Clarys (Eds.), *Biomechanics in Swimming*. Brussels: Université Libre de Bruxelles, 1971.

Shackel, B.: Man-Computer Interaction—The Contribution of the Human Sciences. Ergonomics, *12*, 519–531, 1969.

Simulation. AGARD Conference Proceedings No. 79. London: Technical Editing and Reproduction Ltd., 1971 (AD 717223; AGARD-CP-79-70).

Sukop, J., Petak, K. L., and Nelson, R. C.: An On-Line Computer System for Recording Biomechanical Data. Res. Q. Amer. Assoc. Health Phys. Ed., *42*, 101–102, 1971.

Sutherland, I. E.: Computer Displays. Sci. Am., *222*, 56–81, 1970 (June).

CHAPTER 9

Literature Sources

IT IS GENERALLY ACKNOWLEDGED THAT THE RESEARCHER must be cognizant of the literature in his area of interest. Before embarking upon a research project, it is necessary to ascertain the "state of the art," determine what is known on the subject, what questions have arisen from previous work, which areas require more study and what benefit can be gained from the experience of other investigators. The current explosion of literature (over 100,000 technical journals), however, makes it extremely difficult for the scientist to keep up with the latest information in his field (Selye, 1970).

The literature relevant to biomechanics encompasses several disciplines. Publications in engineering, medicine and physical education must be consulted. In addition, the sophisticated nature of current instrumentation often necessitates reference to technical reports dealing with such topics as photography, computer technology and the measurement and recording of biological parameters. Unless great care is taken, pertinent material may be inadvertently overlooked. Therefore, it is important to identify both the primary sources and the reference literature containing information on current publications which are relevant to the biomechanics of sport.

REFERENCE LITERATURE

Reference sources include abstracts, reviews, indexes and bibliographic citations. Also considered in this category are computerized search services. In most cases, the literature is classified according to subject. Although the term biomechanics has been used in abstract and bibliographic citation publications with increasing frequency since 1968, material on this topic may be found under several different subject headings. These include ergonomics, human factors, human engineering, electromyography, kinesiology and bioengineering. Since subject categories continue to be revised, periodic checks should be made to identify other classifications which may contain information on biomechanics.

Bibliographic Citations. Medical literature relating to biomechanics is cited in the monthly issues of *Index Medicus* under such subjects as biomechanics, biomedical engineering and electromyography. *Index Medicus,* published by the National Library of Medicine, contains references from over 2,200 medically oriented journals throughout the world. It also includes a section devoted to medical reviews in which specific mention is made of articles which are well documented with respect to recent biomedical literature. At the conclusion of each calendar year, subject and author volumes of *Cumulated Index Medicus* are printed.

A somewhat different organizational structure is found in the weekly *Current Contents,* a publication of the Institute for Scientific Information (ISI). Rather than sacrifice time in classifying articles according to subject, the tables of contents of over 4,000 journals are printed. Addresses of the authors are supplied, thus making it possible to request reprints of articles of particular interest. Although *Current Contents* appears in six subject matter areas, the three containing journal contents most related to biomechanics are as follows: *Current Contents/Life Sciences, Current Contents/Physical and Chemical Sciences* and *Current Contents/Engineering and Technology*. Some duplication of journal coverage is encountered in these three sections.

The *Science Citation Index* (SCI), a quarterly publication of ISI, may provide a starting point in the search for pertinent literature. The first step is to identify a major work or an important paper on a specific topic. Reference to the *Citation Index* will then furnish information on who cited the work in a particular calendar year. The citing authors are subsequently located in the *Source Index* to obtain complete bibliographic information on their particular publication. In this way, additional research on the topic in question may be identified.

Among other sources which may be consulted are the *Engineering Index, PANDEX Index to Scientific and Technical Literature, Applied Science & Technology Index,* and the *Government Reports Index. Technical Translations,* a publication of the National Technical Information Service (NTIS), lists foreign language literature which has been translated into English by government and private sources.

Abstracts and Reviews. *Government Reports Announcements* (GRA), formerly *U.S. Government Research and Development Reports,* is a semimonthly NTIS abstract journal which cites reports of research and development made available to the public by such government agencies as the Department of Defense, Atomic Energy Commission and the National Aeronautics and Space Administration. Material is classified according to major subject fields and further subdivided into groups. Entries include the title, bibliographic information, accession number, price and usually an abstract. Among the reports of interest to researchers in biomechanics are those listed in the human factors engineering group within the behavioral and social science field and the bioengineering and bionics groups in the field of biological and medical sciences. NTIS also provides a Fast Announcement Service which alerts the subscriber to the publication of materials in his areas of concern.

International Aerospace Abstracts (IAA) of the American Institute of Aeronautics and Astronautics and *Scientific and Technical Aerospace Reports* (STAR) of the National Aeronautics and Space Administration contain abstracts of world literature in the fields of aeronautics and the science and technology of space. STAR deals with unpublished or report literature including the scientific and technical reports of universities, government agencies and research organizations. Parallel subject coverage is given by IAA to published work in periodicals, conference proceedings and papers of academic and professional organizations. Both IAA and STAR have identical subject categories. Material pertinent to biomechanics is often included in the bioscience and biotechnology sections.

Applied Mechanics Reviews, published each month by the American Society of Mechanical Engineers, contains critical reviews of applied mechanics and related engineering science literature. Relevant subject headings include biomechanics and biotechnology.

For the researcher utilizing the computer for more than the customary data reduction, processing and statistical analysis, various computer abstracting journals may be of value. *Computer and Control Abstracts,* Series C of Science Abstracts, includes references to articles on computer simulation, numerical analysis and graphics systems. *Computing Reviews* provides critical information on computer science publications for computer-oriented persons in other fields while *Computer Abstracts* contains material from conference proceedings, books, periodicals, patents and government reports.

In the realm of medical literature, the Rehabilitation and Physical Medicine section of *Excerpta Medica* abstracts biomechanics articles under these subject headings: anatomy, physiology, kinesiology, biomechanics, physics and engineering, and function tests.

Completed Research in Health, Physical Education and Recreation, published by the American Association for Health, Physical Education and Recreation, summarizes the graduate research in the fields indicated. Canadian and American master's theses and doctoral dissertations are abstracted although the coverage is not comprehensive. Entries are classified according to the degree-granting institution submitting the reports. *Dissertation Abstracts International,* printed by University Microfilms, is divided into humanities (A) and sciences (B) sections. Both author and subject indexes are provided each month and cumulated annually. Since there are no biomechanics, bioengineering or physical education classifications, reference has to be made to other categories such as engineering, physics and physiology. *Masters Abstracts,* which contain summaries of selected master's theses, are also available.

Bibliographies. Complete or selected bibliographies can prove useful in the initial stages of a literature search. These may be found among the references of well-documented kinesiology and biomechanics publications. Bibliographies for specific sports, contained in most *Division for Girls and Women's Sports Guides,* may provide some material related to the biomechanics of sport. Recently completed doctoral dissertations and research reports tend to be among the best bibliographic sources in their area of concentration. Occasionally, reference lists for given subjects are printed for the benefit of other researchers. These are exemplified by Recla (1969) in kinesiology and biomechanics; Powell (1967 and 1970) in track and field; Evans (1967) in physical properties of the skeletal system; the sport medicine bibliography (1970) produced by the National Library of Medicine; and the swimming and diving bibliography of the Council for National Cooperation in Aquatics (1968). The *Bibliographic Index,* a cumulative bibliography of bibliographies, may be helpful in locating similar reference lists. Subject headings such as bio-, human, physical and sports may contain or cross reference bibliographies pertaining to biomechanics.

Computer Search and Retrieval. As evidenced by the numerous indexes and abstract periodicals, the volume of material published each month makes computerized literature storage and retrieval services almost indispensable. The Selective Dissemination of Information (SDI) concept upon which such services are based has become widely accepted by the scientific community. As a result, computerized retrieval systems have begun to proliferate. MEDLARS (Medical Literature Analysis and Retrieval System) of the National Library of Medicine and ASCA IV (Automatic Subject Citation Alert) of ISI are two of the most widely used. A recently instituted NTISearch of reports listed in *Government Reports Index* may prove of value for some aspects of biomechanics research. In addition, a few SDI systems have been developed on a physical education and sport literature base.

Both current awareness and retrospective literature searches can be conducted. In the former case, the user receives computer output periodically on current articles or reports in his specific area of interest. In a retrospective or demand search, on the other hand, the computer scans the literature tapes of the preceding two or three years and provides pertinent bibliographic citations. To obtain best results, an accurate profile of the researcher's needs and interests must be carefully developed. Such a profile is built up from a number of key words and descriptive phrases defining the research problem for which the search is required. Because of the rapid advances in this area, it is advisable to consult a reference librarian about new developments both for assistance in creating an effective profile and for details about best utilizing computerized literature services.

A concerted effort is required to keep abreast of the current literature in biomechanics. Index and abstract journals should be checked periodically to identify pertinent material. To keep unnecessary duplication of effort to a minimum, these publications must be selected to include as many sources as possible while avoiding too much overlap in journal and report coverage. The use of computerized storage and retrieval services, where feasible, will considerably reduce the amount of time which has to be devoted to this phase of research.

PRIMARY SOURCES

Detailed information on the many facets of biomechanics can be found in research reports, theses and dissertations, journal articles, books and conference proceedings in the fields of physical education, medicine and engineering.

Reports. Reports published by government agencies, universities, industry and research foundations furnish in-depth accounts of studies which have been undertaken. They are generally well documented and contain information on instrumentation and techniques as well as the results of the particular investigation. Many of these publications, along with a large number of technical translations, are available in paper copy or microfiche from the NTIS. While the majority of these reports do not deal specifically with sport, general biomechanics topics such as body segment parameters, strength, computer simulation of human movement and techniques for studying body motion are covered.

Theses and Dissertations. Copies of theses and dissertations are customarily retained by the institution granting the degree. In many cases, they may be borrowed for limited periods of time through Interlibrary Loan. All dissertations in physical education, along with selected master's theses, are reproduced on microcards by the School of Health, Physical Education and Recreation at the University of Oregon. A microcard bulletin furnishes a list of those which are available. Dissertations in all areas may be purchased on microfilm from University Microfilms in Ann Arbor, Michigan. Xeroxed copies may also be ordered.

Journals. Since biomechanics is multidisciplinary, journals from several fields are potential sources of relevant literature. Among those which may contain articles pertaining to biomechanics are the following:

Physical Education and Sport

Education Physique et Sport (French)
Exercise and Sport Sciences Reviews
Kinanthropologie (French)
Journal of Health, Physical Education and Recreation
Journal of Sports Medicine and Physical Fitness
Medicine and Science in Sports
Mouvement (French)
Research Quarterly
Sportarzt und Sportmedizin (German)
The Royal Canadian Legion's Coaching Review
Swimming Technique
Theorie und Praxis der Korperkultur (German)
Track Technique
Yessis Translation Review

Human Engineering

Applied Ergonomics

Bio-Medical Engineering
Ergonomics
Human Factors
Journal of Biomechanics
Journal of Industrial Engineering
Le Travail Humain (French)
Medical and Biological Engineering

Physical and Rehabilitation Medicine

American Journal of Physical Medicine
Annals of Physical Medicine
Artificial Limbs
Bulletin of Prosthetics Research
Journal of Bone and Joint Surgery
Journal of the American Medical Association
Physical Therapy Review

Physiology

Acta Physiologica Scandinavica
Internationale Zeitschrift fur angewandte Physiologie einschliesslich Arbeitsphysiologie
Journal of Applied Physiology
Journal of Physiology

Photoinstrumentation

Industrial Photography
Journal of the Society of Motion Picture and Television Engineers
Photogrammetric Engineering
Photographic Applications in Science, Technology and Medicine
Photographic Methods for Industry
Research Film

Mechanics

American Journal of Physics
Experimental Mechanics
Journal of Applied Physics
Physics Today

General and Miscellaneous

American Journal of Physical Anthropology
Annals of the New York Academy of Science
Bulletin of Mathematical Biophysics
Computers in Biology and Medicine
Human Biology
Journal of Psychology
Mathematical Biosciences
Nature
Proceedings of the Royal Society
Research/Development

Science
Science Journal
Scientific American

University libraries subscribe to many of these periodicals. Others can be borrowed through Interlibrary Loan. Reprints of current articles may be requested from the author while those published earlier can be photocopied on inexpensive machines located in most centers.

Conference Proceedings. Symposia devoted to specific areas of biomechanics are being held with increasing frequency. The proceedings of these meetings provide a collection of current research papers and also give an indication of the specific interests of the researchers. Some conferences place their emphasis upon the biomechanics of sport. These are exemplified by the International Seminars on Biomechanics sponsored by the Research Committee of the International Council of Sport and Physical Education (UNESCO) which have been held every two years since 1967 (Wartenweiler *et al.*, 1968; Vredenbregt and Wartenweiler, 1971) and the Biomechanics Symposium at Indiana University (Cooper, 1971). Symposia devoted to other facets of biomechanics such as the First Rock Island Arsenal Biomechanics Symposium (Bootzin and Muffley, 1969), which was highly oriented toward military applications, and the one held in Glasgow, which stressed the significance of biomechanics in medicine (Kenedi, 1965), may contain some papers of interest to sport biomechanists.

Books. A list of many of the books which are related to biomechanics is found at the end of the chapter. Because of the large number, it is not possible to elaborate on each of these publications. They can, however, be categorized as follows: general kinesiology books which are intended for use in undergraduate courses and which contain a substantial portion of applied anatomy; those devoted to biomechanics or the biomechanics of sport in general; books on basic mechanics, usually from a physics or engineering series; and those dealing with the application of mechanical principles to specific sports. At present, there is a need for more publications in the latter category which are based upon sound research evidence and practical knowledge of the activity.

SUMMARY

As can be appreciated from the previous discussion, the literature in biomechanics is extensive and overlaps several disciplines. Although it is not an easy task to keep up to date on the current publications in one's area of interest, it is a necessary one if duplication of effort is to be minimized. A knowledge of abstracting publications, indexes, computer search systems, bibliographies and relevant journals should facilitate the literature search aspect of biomechanics of sport investigations.

SELECTED REFERENCES

Abendschein, W. F.: Orthopaedics and the Information Explosion. J. Bone Joint Surg. *52A*, 184–193, 1970.

Alt, F. (Ed.): *Advances in Bioengineering and Instrumentation*. Vol. 1. New York: Plenum Press, 1966.

Amar, J.: *The Human Motor*. Dubuque, Iowa: Brown Reprints, 1972.

Anderson, J. M.: *Human Kinetics and Analyzing Body Movements*. London: Heinemann, 1951.

American Society of Mechanical Engineers: *Biomechanics Monograph*. New York: ASME, 1967.

American Society of Mechanical Engineers: Biomechanical and Human Factors Symposium. New York: ASME, 1967.

Bade, E.: *The Mechanics of Sport*. Kingswood, Surrey: Glade House, 1962.

Ballreich, R.: *Weitsprung-Analyse*. Sportwissenschaftliche Arbeiten Band 3. Berlin: Bartels & Wernitz, K. G., 1969.

Ballreich, R.: Weg- und Zeitmerkmale von Sprintbewegungen. Sportwissenschaftliche Arbeiten Band 1. Berlin: Bartels & Wernitz, 1969.

Basmajian, J. V.: *Muscles Alive—Their Functions Revealed by Electromyography*. 2nd Ed., Baltimore: Williams & Wilkins, 1967.

Bernstein, N. A.: *The Co-ordination and Regulation of Movements*. New York: Pergamon Press, 1967.

Bootzin, D., and Muffley, H. C.: *Biomechanics*. New York: Plenum Press, 1969.

Broer, M.: *An Introduction to Kinesiology*. Englewood Cliffs, N.J.: Prentice-Hall, 1968.

Broer, M.: *Efficiency of Human Movement*. 2nd Ed., Philadelphia: W. B. Saunders, 1966.

Broer, M. R., and Houtz, S. J.: *Patterns of Muscular Activity in Selected Sport Skills—An Electromyographic Study*. Springfield, Ill: C. C Thomas, 1967.

Brown, J. E.: The CAN/SDI Project. Spec. Libr., *60*, 501–509, 1969.

Brunnstrom, S.: *Clinical Kinesiology*. 2nd Ed., Philadelphia: F. A. Davis, 1966.

Bunn, J. W.: *Scientific Principles of Coaching*. 2nd Ed., Englewood Cliffs, N.J.: Prentice-Hall, 1972.

Carlsöö, S.: *Manniskans Roresler*. Stockholm: Personaladministrativa Radet, 1968.

Cooper, J. M., and Glassow, R. B.: *Kinesiology*. 3rd Ed., St. Louis: C. V. Mosby, 1972.

Cooper, J. M. (Ed.): *Selected Topics on Biomechanics*. Chicago: Athletic Institute, 1971.

Council for National Cooperation in Aquatics: *Swimming and Diving: A Bibliography*. New York: Association Press, 1968.

Donskoi, D. D.: *Biomechanik der Korperübungen*. Berlin: Sportverlag, 1961.

Ducroquet, R., Ducroquet, J., and Ducroquet, P.: *Walking and Limping—A Study of Normal and Pathological Walking*. Philadelphia: Lippincott, 1968.

Duvall, E. N.: *Kinesiology, The Anatomy of Motion*. Englewood Cliffs, N.J.: Prentice-Hall, 1959.

Dyson, G.: *The Mechanics of Athletics*. 5th Ed., London: University of London Press, 1970.

Ecker, T.: *Track and Field Dynamics*. Los Altos, Calif.: Tafnews Press, 1971.

Evans, F. G.: Bibliography on the Physical Properties of the Skeletal System. Artif. Limbs, *11*, 48–66, 1967 (Autumn).

Evans, F. G. (Ed.): *Biomechanical Studies of the Musculo-Skeletal System*. Springfield, Ill.: C. C Thomas, 1961.

Fetz, F., and Opavsky, P.: *Biomechanik des Turnens*. Frankfurt: Limpert, 1968.

Frankel, V. H., and Burstein, A. H.: *Orthopaedic Biomechanics: The Application of Engineering to the Musculoskeletal System*. Philadelphia: Lea & Febiger, 1970.

Frost, H. M.: *An Introduction to Biomechanics*. Springfield, Ill.: C. C Thomas, 1967.

Fung, Y. C., Perrone, N., and Anliker, M.: *Biomechanics—Its Foundations and Objectives*. Englewood Cliffs, N.J.: Prentice-Hall, 1972.

Fung, Y. C. (Ed.): *Biomechanics*. New York: ASME, 1966.

Govaerts, A.: *La Biomecanique—Nouvelle Methode d'Analyse du Mouvement*. Brussels: Presses Universitaires de Bruxelles, 1971.

Harre, D.: *Trainingslehre Einfuhrung in die allgemeine Trainingsmethodik*. Berlin: Sportverlag, 1970.

Hawley, G.: *An Anatomical Analysis of Sport*. New York: Barnes, 1940.

Hay, J. G.: *The Biomechanics of Sports Techniques*. Englewood Cliffs, N.J.: Prentice-Hall, 1973.

Hochmuth, G.: *Biomechanik sportlicher Bewegungen*. Berlin: Sportverlag, 1967.

Hochmuth, G.: *Biomechanik*. Leipzig: Deutsche Hochschule fur Korperkultur, 1962.

Hopper, B. J.: *Notes on the Dynamical Basis of Physical Movement*. Strawberry Hill Booklets No. 2. Liverpool: Kilburns (Printers), 1959.

Houghton, B.: *Computer Based Information Retrieval Systems*. Hamden, Conn.: Archon Books, 1969.

Jensen, C. R., and Schultz, G. W.: *Applied Kinesiology*. New York: McGraw-Hill, 1970.

Joseph, J.: *Man's Posture—Electromyographic Studies*. Springfield, Ill.: C. C Thomas, 1960.

Kelley, D. L.: *Kinesiology—Fundamentals of Motion Description*. Englewood Cliffs, N.J.: Prentice-Hall, 1971.

Kenedi, R. M. (Ed.): *Biomechanics and Related Bioengineering Topics*. New York: Pergamon Press, 1965.

Kenyon, G. K.: The Retrievel of Information. PE 990, University of Wisconsin, 1968 (mimeographed).

Kinesiology 1971. Washington: AAHPER, 1971.

Kinesiology 1968. Washington: AAHPER, 1968.

Krogman, W. M., and Johnston, F. E. (Eds.): *Human Mechanics—Four Monographs Abridged*. Wright-Patterson Air Force Base, Ohio, 1963 (AMRL-TDR-63-123; AD 600 618).

Kuhlow, A.: *Analyse moderner Hochsprungtechniken*. Sportwissenschaftliche Arbeiten Band 5. Berlin: Bartels & Wernitz KG, 1971.

Lewillie, L., and Clarys, J. P. (Eds.): *Biomechanics in Swimming*. Brussels: Universite Libre de Bruxelles, 1971.

Lipovetz, F. J.: *Basic Kinesiology*. Minneapolis: Burgess, 1952.

Logan, G. A., and McKinney, W. C.: *Kinesiology*. Dubuque, Iowa: Wm. C. Brown, 1970.

MacConaill, M. A., and Basmajian, J. V.: *Muscles and Movements—A Basis for Human Kinesiology*. Baltimore: Williams & Wilkins, 1969.

McCloy, C. H.: The Mechanical Analysis of Motor Skills. In W. R. Johnson (Ed.), *Science and Medicine of Exercise and Sports*. New York: Harper, 1960.

Marey, E. J.: *Movement*. London: Neineman, 1895.

Marinacci, A. A.: *Applied Electromyography*. Philadelphia: Lea & Febiger, 1968.

Novak, A.: *Biomechanika Telesnych Cviceni*. Prague: Statni Pedagogicke Nakledatelstvi, 1965.

O'Connell, A. L., and Gardner, E. B.: *Understanding the Scientific Basis of Human Movement*. Baltimore: Williams & Wilkins, 1972.

Plagenhoef, S.: *Patterns of Human Motion—A Cinematographic Analysis*. Englewood Cliffs, N.J.: Prentice-Hall, 1971.

Powell, J. T.: A Compilation and Analysis of Classified, Indexed and/or Completed Research in Track and Field Athletics. Coaching Rev., *7*, 8–11, 1970 (March) and *5*, 7–11, 1967 (June).

Rasch, P. J., and Burke, R. K.: *Kinesiology and Applied Anatomy*. 4th Ed., Philadelphia: Lea & Febiger, 1971.

Recla, J.: Die Biomechanik der Leibesübungen in der internationalen Literatur der Gegenwart. Institut für Leibeserziehung der Universität Graz, 1969.

Rogers, E. M.: *Physics for the Inquiring Mind*. Princeton: Princeton University Press, 1960.

Scott, M. G.: *Analysis of Human Motion*. 2nd Ed., New York: Appleton-Century-Crofts, 1963.

Selinger, V., and Novak, A.: *Biomechanika Sportovniho Pohybo*. Prague: Sportovni a Turisticke Nakladatelstvi, 1960.

Selye, H.: Can We Cope with the 'Literature Explosion'? J. Exp. Clin. Med., *1*, 3–8, 1970.

Steindler, A.: *Kinesiology of the Human Body under Normal and Pathological Conditions*. Springfield, Ill.: C. C Thomas, 1955.

Steindler, A.: *Mechanics of Normal and Pathological Locomotion in Man*. Springfield, Ill.: C. C Thomas, 1935.

Symposium Theorie der Sporttechnik. Warsaw: Akademie fur Korpererziehung, 1968.

Thompson, C. W.: *Manual of Structural Kinesiology*. 6th Ed., St. Louis: C. V. Mosby, 1969.

Tricker, R. A. R., and Tricker, B. J. K.: *The Science of Movement.* New York: American Elsevier, 1967.

Vredenbregt, J., and Wartenweiler, J. (Eds.): *Biomechanics II.* Baltimore: University Park Press, 1971.

Wartenweiler, J., Jokl, E., and Hebbelinck, M. (Eds.): *Biomechanics.* Baltimore: University Park Press, 1968.

Weinstock, M.: Citation Indexes. Encycl. Lib. Info. Sci., *5*, 16–40, 1971.

Wells, K. F.: *Kinesiology.* 5th Ed., Philadelphia: W. B. Saunders, 1971.

Wickstrom, R. L.: *Fundamental Motor Patterns.* Philadelphia: Lea & Febiger, 1970.

Williams, M., and Lissner, H. R.: *Biomechanics of Human Motion.* Philadelphia: W. B. Saunders, 1962.

Winchell, C. M.: *Guide to Reference Books.* 8th Ed., Chicago: American Library Association, 1967.

Wolters, P. H., and Brown, J. E.: CAN/SDI System: User Reaction to a Computerized Information Retrieval System for Canadian Scientists and Technologists. Can. Libr. J., *28*, 20–23, 1971.

Yamada, H.: *Strength of Biological Materials.* F. G. Evans (Ed.). Baltimore: Williams & Wilkins, 1970.

CHAPTER 10

Research Fundamentals in Biomechanics of Sport

RESEARCH, an essential element in the emergence of biomechanics as a viable and mature discipline, is receiving increased attention at fundamental, practical and theoretical levels. Such on-going research programs require careful planning, implementation and evaluation if they are to make a significant contribution to this rapidly developing speciality. While general information on the conduct of research is available in existing publications, there is a need to consider requirements specifically related to the investigation of problems in the biomechanics of sport. Therefore, the discussion presented is intended to meet this need and to assist those interested in carrying out effective research in the mechanics of human motion.

PLANNING

The first step in biomechanics research is the identification of the particular problem to be studied. Personal experience in coaching, teaching and/or research as well as familiarity with the literature on human motion will reveal a number of questions requiring scientific investigation. The proposed study may then be classified as theoretical, fundamental or practical in nature, depending upon the ultimate purpose for which the results are intended. Once this level has been determined, a clear statement of the purpose, experimental hypotheses and scope of the investigation must be prepared. Limiting factors must be identified and the overall experimental design delineated. Consideration has to be given to the number and type of subjects to be chosen as well as to their method of selection. The number of test sessions, experimental conditions and the sequence of individual trials must be planned. The required equipment and an adequate testing site need to be reserved well in advance.

PILOT STUDY

Prior to launching the main investigation, it is advisable to conduct a preliminary or pilot project. This is normally done with a limited number of subjects in an attempt to simulate the procedures of the actual study. The testing environment should provide as nearly-normal circumstances for the subjects as possible. When studying sports skills at the practical level, a choice may have to be made between a less consistent but more natural sports environment and a better-controlled, although artificial, laboratory setting. The researcher must be alert to factors such as bright lighting, unfamiliar laboratory instruments, noise due to the operation of equipment and the presence of research assistants which may affect the performance of the subjects. Data recording and analysis procedures including the selection of biomechanical variables, data collection, development of computer programs, completion of calculations and preliminary data analysis should be detailed. Training research assistants during the pilot investigation will help to ensure standardization and consistency during the actual experiment. As a result of the preliminary project, it will be possible to identify and eliminate potential sources of difficulty; streamline the procedures; estimate the time required for each phase of the actual study, the cost of the total project and the number of personnel required.

MAIN STUDY

The detailed plans for the main investigation are based upon the experience gained during the pilot study. Consideration must be given to documentation, instrumentation, potential sources of experimental error, statistical analysis, dissemination of results and evaluation.

Documentation. Specific written instructions for each phase of the study must be prepared so that the planned procedures are followed precisely. A diary should be kept during the conduct of the study to record routine details and any unusual occurrences. Such information may provide insight for future experimentation and will be needed for the preparation of the written report on the project. It is recommended that still photographs of the test area, equipment and subjects be taken while the study is underway. Both color slides to be used for presentation of the research results at professional meetings and black and white prints for inclusion in published papers are required.

Instrumentation. Recent advances in instrumentation systems have contributed greatly to the quality and quantity of research in the biomechanics of sport. Before embarking upon a research project, the investigator should make a thorough evaluation of available measurement systems. Consultation with electronic and/or photographic personnel is advisable to determine whether construction or purchase of new equipment or modification of that which is already available will best meet the requirements of the proposed study. Consideration must also be given to cost, accuracy and consistency of operation, ease of data recording and processing, compatibility of measuring and recording components, sampling rate, simplicity of operation and maintenance, portability if required and the applicability of the equipment to related studies.

Once the proper instrumentation is chosen, the researcher has to become completely familiar with its operation. Calibration of the equipment must be performed regularly during the conduct of the experiment to ensure that it is functioning properly. This applies to such instruments as force transducers, cameras, accelerometers, strobe lights and electrogoniometers. Availability of performance curves supplied by the manufacturer does not relieve the researcher of the responsibility for assessing the accuracy and consistency of an instrument.

Experimental Error. It is important to identify and isolate the factors which contribute to experimental error. For example, regular calibration of equipment assures the researcher that instrument error has been minimized. The testing environment itself may contribute to variability in performance if standardized conditions are not maintained. This implies that careful records must be kept of the exact location of test apparatus so that it can be repositioned if necessary. Body markings on subjects must be placed consistently and accurately from one test day to the other. Standardization is especially critical in longitudinal studies since they require several repetitions of the same test conditions at set time intervals. The possibility that the data may be unintentionally biased by the experimenter must be guarded against. This can occur subtly during testing sessions, data analysis and even in the interpretation of results. A more consistent test situation can be provided if the number of researchers in direct contact with the subjects is kept to a minimum. The same principle applies to analyzing the data, making measurements from film, recording and card punching data, and performing statistical analyses. If transcription is absolutely necessary, adequate double checking procedures must be employed to ensure that the data are copied correctly. It is important that assistants participating in various stages of the research be thoroughly trained and their objectivity evaluated periodically.

Variability in the performance of the subjects can be expected in most biomechanical studies. In fact, the characteristic of variability is often of prime importance in assessing the biomechanical components of sports performance. In many instances it is important to study more than one trial on a given day and also to repeat the test sequence on one or more subsequent days. A better estimate of the "normal" performance can be secured if the movement is replicated in the experiment.

STATISTICAL ANALYSIS

Research in most disciplines has been markedly influenced by the emphasis upon inferential statistical treatment of data. Physical education has also been affected by this trend, as is evident from a cursory examination of the *Research Quarterly*, the principal American research journal in this field. The sport biomechanist must develop a good working knowledge of statistics to be able to read and understand the literature and to utilize this tool in his own research. Statistical treatment of the data serves primarily to aid the researcher in his interpretation of the results, and does not eliminate the need for thorough understanding of the data and their practical significance.

Availability of computer library programs for most statistical tests has greatly reduced the time required to complete the computational phase. However, the utilization of computer methods is ill-advised unless the researcher understands fully the mathematical manipulations being performed. When large quantities of data are being processed, some of the computations should be duplicated by hand calculation to make certain that the computer results are accurate. Errors can occur when input instructions are not followed precisely or when raw data are inaccurately presented or misread. Prior to performing the statistical analyses, it is recommended that the data be compared with those reported in the literature and be scrutinized from a descriptive standpoint. Hand plotting of group or individual mean values with respect to training session or some other factor gives the investigator some "feeling" for these data. The inferential statistical tests merely serve to establish the probability of the occurrence of various events and lend quantitative credence to conclusions which are drawn.

Selection of the statistical procedures to be employed is influenced by the number of subjects to be utilized in the investigation. Studies involving a few subjects are best suited to a case study approach in which individuals are examined separately. This would occur, for example, in a longitudinal study of the movement patterns of highly skilled athletes or young children. If small groups of subjects are used and inferences are to be made, then nonparametric statistical methods would be appropriate. These tests, referred to as distribution-free or ranking tests, are described in texts such as the one by Siegel (1956). The main advantage lies in their usefulness with small samples in which no assumptions need be made about the distribution from which the scores are taken. In studies involving groups of subjects, a more complex experimental design is required. The variability due to order, sequence and training effects in addition to that caused by individuals must be identified and taken into account. These factors should be considered in the establishment of the test protocol for the experiment since the appropriate statistical treatment is dependent upon the experimental design. Detailed descriptions of the various designs and multiple-group statistical tests are available in Steel and Torrie (1960), Winer (1972) and Runyon and Haber (1971).

Dissemination of Results. The results of biomechanics research can be utilized in a variety of ways. In a practical study dealing with a few skilled athletes, the findings may be interpreted to the teacher or coach who, in turn, can apply them to improve the performance of these individuals. Results of a practical study involving a larger number of subjects may have generalized implications, and therefore, should be disseminated by publication in coaching and sport technique journals as well as through presentations at workshops and clinics. Fundamental and theoretical work is normally reported at scientific meetings and in research journals. Detailed suggestions for planning and presenting a paper have been published by Williams (1965). Since it is a professional responsibility to share quality work with other researchers, papers should be submitted to publications read by those concerned with that particular aspect of human motion. In many instances, a second paper covering the same material may be written in nontechnical language and submitted to a professional or popular journal.

Careful consideration must be given to the selection of the appropriate journal and preparation of the manuscript. The study must be compatible with the general theme of the publication or it will very likely be rejected. Strict adherence to the editorial guidelines is required. These are included periodically in the journal or may be secured from the editorial office. Care must be taken that the writing style meets the standards of scientific publications. The specific suggestions for the preparation of a paper made by Smith and his associates (1971) may be of considerable value. Prior to preparation of the final copies, it is highly recommended that two or three colleagues critically review the manuscript. Also the final draft should be proofread by at least two persons to make certain that all typographical errors are eliminated. The required number of copies of the original may then be made using a quality photocopier. It is advisable to have graphical illustrations such as block diagrams and bar graphs prepared by a professional draftsman. A sufficient number of illustrations and photocopies should be included so that each submitted copy of the manuscript is complete. Labels containing the author's name and title of paper should be secured to the back of illustrations and photographs to aid in identification in case they become detached from the manuscript. A duplicate of the submitted manuscript including photos and illustrations must be retained by the authors. If the paper has scientific merit and meets the editorial standards, it will normally be accepted for publication although some changes may be required. Before the article appears in print the author receives page or galley proofs. This represents the last opportunity to make certain that no errors are present and, consequently, thorough scrutiny of the text, tables and references is imperative.

Regardless of the level of research, it is the investigator's responsibility to make his findings available to others. The process of preparing and submitting a manuscript to be considered for publication provides an important means of judging the quality of the work. Since the review of papers is normally done by experienced investigators, their constructive comments can be of significant value. Through the process of presenting papers, publishing research results and interacting professionally with persons of similar interests, the sport biomechanist becomes a member of a larger group of devoted professionals.

Evaluation. Establishment of an on-going research program is dependent upon effective evaluation of each individual research project. The most productive work results from a systematic progression of studies, each of which builds on the results and experience of the preceding investigation. Consequently, it is necessary to document carefully any difficulties encountered and to review the procedures and results in anticipation of planning the next investigation.

In the interest of effective personal and public relations, it is recommended that letters of appreciation be sent to the subjects and other persons such as teachers or coaches who cooperated in the study. When appropriate, the subjects may receive a summary of the study and specific information about their performance. Special assistance from cooperating medical, technical or administrative personnel needs to be acknowledged and appreciation extended to members of the research team.

SUMMARY

Biomechanics research at the practical, fundamental and theoretical levels has increased markedly in recent years. As this work becomes more complex and sophisticated, it is necessary for researchers to improve their instrumentation systems, experimental procedures, data recording, analysis methods and means of disseminating their findings. The ultimate success of biomechanics research in sport will depend greatly upon the creativity and competence of the researchers in their continual efforts to strive for more effective means of investigating human movement.

SELECTED REFERENCES

AAHPER: *Research Methods in Health, Physical Education, and Recreation*. Washington: AAHPER, 1959.

AAHPER: Research Quarterly Evaluative Criteria. Res. Q. Amer. Assoc. Health Phys. Ed., *43*, 253–255, 1972.

Clarke, D. H., and Clarke, H. H.: *Research Processes in Physical Education, Recreation and Health*. Englewood Cliffs, N.J.: Prentice-Hall, 1970.

Myers, J. L.: *Fundamentals of Experimental Design*. 2nd Ed., Boston, Mass.: Allyn & Bacon, 1972.

Runyon, R. P., and Haber, A.: *Fundamentals of Behavioral Statistics*. 2nd Ed., Reading, Mass.: Addison-Wesley, 1971.

Siegel, S.: *Nonparametric Statistics for the Behavioral Sciences*. New York: McGraw-Hill, 1956.

Smith, O. W., Smith, P. C., Scheffers, J., and Steinmann, D.: Common Errors in Reports of Psychological Studies. Res. Q. Amer. Assoc. Health Phys. Ed., *42*, 466–470, 1971.

Steel, R. G. D., and Torrie, J. H.: *Principles and Procedures of Statistics*. New York: McGraw-Hill, 1960.

Williams, P. C.: Suggestions for Speakers and Standards for Slides. Institute of Biology Journal, 1–8, 1965 (May).

Winer, B. J.: *Statistical Principles in Experimental Design*. 2nd Ed., New York: McGraw-Hill, 1972.

CHAPTER 11

Development of a Biomechanics Laboratory for Use in Research and Graduate Programs

A NATURAL outgrowth of the increased interest in the biomechanics of sport has been the development of laboratories as part of research and graduate programs. Since the funds available are often limited, maximum utilization of these resources must be achieved. Factors which influence planning decisions are the philosophy of the research programs, utilization of space, personnel, equipment and sources of funds.

PHILOSOPHY

The formulation of a guiding philosophy of the overall research program is paramount since it underlies all other aspects of program development. Many factors must be considered, the first being the role of biomechanics in the research and graduate programs of the department. Will biomechanics be an area of specialization within the graduate program or will it be primarily a faculty research specialty? If graduate education is to be included, it will influence the plans for space utilization and equipment selection. Further, the number and competencies of the faculty and technical staff will also be affected.

Because biomechanics of sport is a relatively broad field it is unlikely that each laboratory will have sufficient resources to conduct research in all subareas. Hence the scope of the program must be limited. This is normally done on the basis of the research interests and competencies of the faculty member(s). For example, one group may decide to concentrate on the study of the external mechanics of human movement using cinematographic methods while another may choose to specialize in the study of fundamental movements utilizing bioelectronic and electromyographic techniques. Concentration in such a subarea affects the space and equipment requirements as well as the selection of graduate student candidates.

The focus of the program may also be influenced by the available opportunities for interdisciplinary activities. Researchers in fields of mechanical engineering, anatomy, physiology, biophysics, computer science, physiotherapy and physiological psychology may be interested in cooperative ventures. Interaction with specialists in these fields is of primary importance to the vigor and strength of a developing program since it makes possible the exchange of ideas, technical assistance and advice. However, this is sometimes difficult since research in physical education is not often understood by persons in other disciplines. Experience has shown, however, that excellent working relations can be developed upon a feeling of mutual respect.

Involvement of graduate students in these interdisciplinary activities is of paramount importance to their education. The basic knowledge upon which the study of biomechanics depends is found in a number of disciplines. Consequently, it is impossible to provide all of the necessary course content within a graduate curriculum in physical education. Graduate students *must* take courses in a number of related areas if they expect to be adequately prepared for future careers in biomechanics research. The specific courses taken will be influenced by the proposed area of concentration and the availability of courses within the university. Students specializing in the mechanics of motion should have adequate quantitative skills and supporting course work in mathematics. Additional courses should be selected from fields such as engineering mechanics, physics, bioengineering, computer science and biophysics. Advice on course selections can be secured from faculty members in these disciplines.

A quality doctoral program includes a variety of experiences designed to develop the research capabilities of the students. Experience in an on-going research program throughout the training period is essential. Extensive practical experience in the care and operation of laboratory equipment must be provided. Students should participate in all phases of research projects including planning, data collection and analysis, preparation of the manuscripts for publication and presentation of papers at professional meetings. They should assist in the preparation of grant proposals to become acquainted with guidelines, budget preparation and procedures for transmitting the proposal. Conduct of independent research projects (prior to their dissertation research) should be encouraged. Experience in teaching undergraduate courses in biomechanics is highly desirable and, when possible, doctoral students should assist in advising master's degree candidates. Providing these experiences in the doctoral program gives greater assurance that the graduates will be able to assume positions of responsibility and actively begin their professional careers.

SPACE UTILIZATION

Under normal circumstances, special facilities, although desirable, are not essential for biomechanics research. It is often necessary for persons initiating a laboratory program to make use of whatever office and work space is available. Modified classrooms as well as locker, equipment and even shower rooms have been successfully used for research purposes.

Because only limited space may be available, it is essential that multiple usage be considered. One area may serve as a student laboratory at times and be used for data collection on other occasions. Since flexibility must also be maintained it is ill-advised to reserve a room for only one function and permit it to sit idle for long periods of time. Efficient scheduling is especially necessary when a number of research projects are being conducted concurrently. The nature of sport biomechanics makes it possible to utilize athletic facilities for filming research projects. Studies on running and hurdling can often (and perhaps should) be conducted on the track. Other sports such as basketball, gymnastics and fencing can be more effectively filmed in a gymnasium or special sports room. An indoor fieldhouse offers the possibility of studying movements normally performed outdoors. Filming of swimmers is facilitated by underwater windows while an elevated platform makes possible the study of sports skills from an overhead view. When possible, 220V electrical outlets should be installed to insure that adequate lighting is available. Although using these facilities results in the necessity of moving the equipment and other inconveniences, it does greatly expand the space available for conducting biomechanics research and allows the subjects to perform in their normal sports environment.

PERSONNEL

Whereas special facilities may not be essential, availability of technical assistance is imperative to the development of a laboratory. Ideally, a full-time electrical engineer should be a member of the staff. He can advise on equipment selection, construct components when it is feasible, design and create new instrumentation systems and participate as a member of the research team. Special training in computer science is desirable since it has become an integral part of biomechanics research. The availability of expert technical assistance is of considerable importance in the development and maintenance of a vital, productive program.

If a full-time position for an engineer cannot be justified then a part-time position may be adequate, especially during the early stages of development. In some universities a "pool" of electronics technicians is available or a contract for service can be arranged with a private firm. The hiring of an electrical engineering student on a graduate assistantship in physical education may offer an alternate solution.

In addition to electronics assistance, it is desirable to have technical expertise in photography and computer technology, depending upon the specific requirements of the program. It is possible that graduate students with special abilities will be able to provide assistance in these areas. Certainly all graduate students in biomechanics must have at least a working knowledge of computer programming. Photographic assistance may be secured from a central photographic laboratory or the audiovisual department found on most campuses.

EQUIPMENT

The quality of research in biomechanics is dependent to a great extent upon the equipment available. A well-equipped laboratory requires a considerable financial outlay. Since funds are usually limited, it is essential that they be used wisely. Items may often be borrowed from other departments or from government or industrial laboratories. This makes it possible to initiate a research program with limited resources and at the same time gain experience in the use of the equipment which may be purchased in the future.

Assuming funds are available, the type of equipment needed should be clearly defined. A choice must then be made. The first step is to identify all companies marketing the desired item and request descriptive brochures. The list of companies published by AAHPER (1971) is of considerable value in locating the manufacturers of specific items. With the advice of an engineer, a comparison should be made based on the price, characteristics and related factors. Sales representatives should be invited to demonstrate the equipment and, if possible, leave a demonstration model for a trial period. When feasible, consideration should be given to the purchase of such a model as this usually results in a financial saving. Availability of service is another important consideration in selecting the specific model. Sales representatives will normally provide names and addresses of persons who have purchased their equipment and can be contacted concerning its quality and performance.

As more equipment becomes available the problem of compatibility arises. If electronic components are to be interconnected to form an instrumentation system, it is necessary that their fundamental characteristics be quite similar. If this is not assured, costly modifications may be required.

It is advisable to maintain a current list of needed equipment so that if funds become available on short notice it will be possible to order equipment quickly. Price quotations from companies should be secured in advance to expedite handling the purchase.

SOURCES OF FUNDS

The principal deterrent to acquiring research equipment is the lack of adequate funds. This is an especially difficult problem for the sports biomechanist because of the general shortage of financial support for research in the field of physical education. Most universities, however, do have resources for support of faculty research which are usually provided by federal institutional grants, state appropriations, foundations and industries. It is essential to seek out these sources and make your needs known. Equipment may be secured by including it as part of a grant proposal to support a specific research project. If the equipment can also be justified for teaching purposes, it may be possible to secure the necessary funds through an instructional budget.

SUMMARY

As the number of biomechanics graduate and research programs increases, it is desirable that this growth proceed in an expeditious manner. Funds to support these programs will continue to be limited so that prudent planning is essential. Suggestions presented in this chapter should serve as guiding principles for persons contemplating development of graduate and research programs in the biomechanics of sport.

SELECTED REFERENCES

AAHPER: Sources of Research Laboratory Equipment. Res. Q. Amer. Assoc. Health Phys. Ed., *42*, 338–353, 1971.

Proceedings: Biomechanics Conference. Pennsylvania State University, 1971.

APPENDIX A

Trigonometry

TRIGONOMETRY, the branch of mathematics concerned with triangle measurement, developed from observations that the ratio of any pair of sides of a right triangle is constant if the size of the acute angle remains fixed. These ratios, called trigonometric functions, are as follows:

$$\textit{sine}\ A \text{ or } \textit{sin}\ A = \frac{\text{side opposite angle } A}{\text{hypoteneuse}}$$

$$\textit{cosine}\ A \text{ or } \textit{cos}\ A = \frac{\text{side adjacent to angle } A}{\text{hypoteneuse}}$$

$$\textit{tangent}\ A \text{ or } \textit{tan}\ A = \frac{\text{side opposite angle } A}{\text{side adjacent to angle } A}$$

$$\textit{cosecant}\ A \text{ or } \textit{csc}\ A = \frac{\text{hypoteneuse}}{\text{side opposite angle } A}$$

$$\textit{secant}\ A \text{ or } \textit{sec}\ A = \frac{\text{hypoteneuse}}{\text{side adjacent to angle } A}$$

$$\textit{cotangent}\ A \text{ or } \textit{cot}\ A = \frac{\text{side adjacent to angle } A}{\text{side opposite angle } A}$$

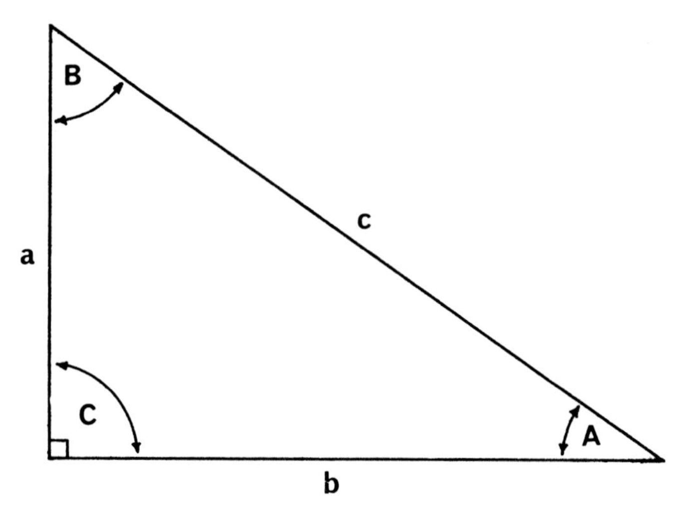

FIGURE A-1. *Right Triangle.*

in which *A* is one of the acute angles of a right triangle and the hypoteneuse, which is the longest side, is always opposite the right angle. With reference to Figure A-1:

$$\sin A = \frac{a}{c} \qquad \cos A = \frac{b}{c} \qquad \tan A = \frac{a}{b}$$

$$\cos B = \frac{a}{c} \qquad \sin B = \frac{b}{c} \qquad \tan B = \frac{b}{a}\ .$$

Two angles are said to be complementary if their sum equals 90 degrees. Since the sum of three internal angles of a triangle is always 180 degrees and since angle *C* is a right angle, then angles *A* and *B* are complementary. From the examples, it can be seen that the cosine of an angle is equal to the sine of its complement, a fact used in determining sine and cosine values on a slide rule.

Since the six trigonometric functions are constants for any given angle regardless of the size of the right triangle, they can be included in table form (Table A-1). Because the hypoteneuse is the longest side of the triangle, the numerical values of sine and cosine must fall between +1.00 and −1.00. Tangents, however, are not so restricted. Figure A-2 shows the sine and cosine functions in graphic form. Table A-2 is presented to assist in reducing the trigonometric functions of obtuse angles to their equivalent acute value. Thus, $\cos(180 - \alpha) = -\cos\alpha$ and $\sin(90 + \alpha) = \cos\alpha$ in which α is an angle less than 90 degrees.

Table A-1. Trigonometric Functions*

Degrees	*Sines*	*Cosines*	*Tangents*	*Cotangents*	
0	.0000	1.0000	.0000		90
1	.0175	.9998	.0175	57.290	89
2	.0349	.9994	.0349	28.636	88
3	.0523	.9986	.0524	19.081	87
4	.0698	.9976	.0699	14.301	86
5	.0872	.9962	.0875	11.430	85
6	.1045	.9945	.1051	9.5144	84
7	.1219	.9925	.1228	8.1443	83
8	.1392	.9903	.1405	7.1154	82
9	.1564	.9877	.1584	6.3138	81
10	.1736	.9848	.1763	5.6713	80
11	.1908	.9816	.1944	5.1446	79
12	.2079	.9781	.2126	4.7046	78
13	.2250	.9744	.2309	4.3315	77
14	.2419	.9703	.2493	4.0108	76
15	.2588	.9659	.2679	3.7321	75
16	.2756	.9613	.2867	3.4874	74
17	.2924	.9563	.3057	3.2709	73
18	.3090	.9511	.3249	3.0777	72
19	.3256	.9455	.3443	2.9042	71
20	.3420	.9397	.3640	2.7475	70
21	.3584	.9336	.3839	2.6051	69
22	.3746	.9272	.4040	2.4751	68
23	.3907	.9205	.4245	2.3559	67
24	.4067	.9135	.4452	2.2460	66
25	.4226	.9063	.4663	2.1445	65
26	.4384	.8988	.4877	2.0503	64
27	.4540	.8910	.5095	1.9626	63
28	.4695	.8829	.5317	1.8807	62
29	.4848	.8746	.5543	1.8040	61
30	.5000	.8660	.5774	1.7321	60
31	.5150	.8572	.6009	1.6643	59
32	.5299	.8480	.6249	1.6003	58
33	.5446	.8387	.6494	1.5399	57
34	.5592	.8290	.6745	1.4826	56
35	.5736	.8192	.7002	1.4281	55
36	.5878	.8090	.7265	1.3764	54
37	.6018	.7986	.7536	1.3270	53
38	.6157	.7880	.7813	1.2799	52
39	.6293	.7771	.8098	1.2349	51
40	.6428	.7660	.8391	1.1918	50
41	.6561	.7547	.8693	1.1504	49
42	.6691	.7431	.9004	1.1106	48
43	.6820	.7314	.9325	1.0724	47
44	.6947	.7193	.9657	1.0355	46
45	.7071	.7071	1.0000	1.0000	45
	Cosines	*Sines*	*Cotangents*	*Tangents*	*Degrees*

*With angles above 45° be sure to use the headings that appear at the *bottom* of the columns.

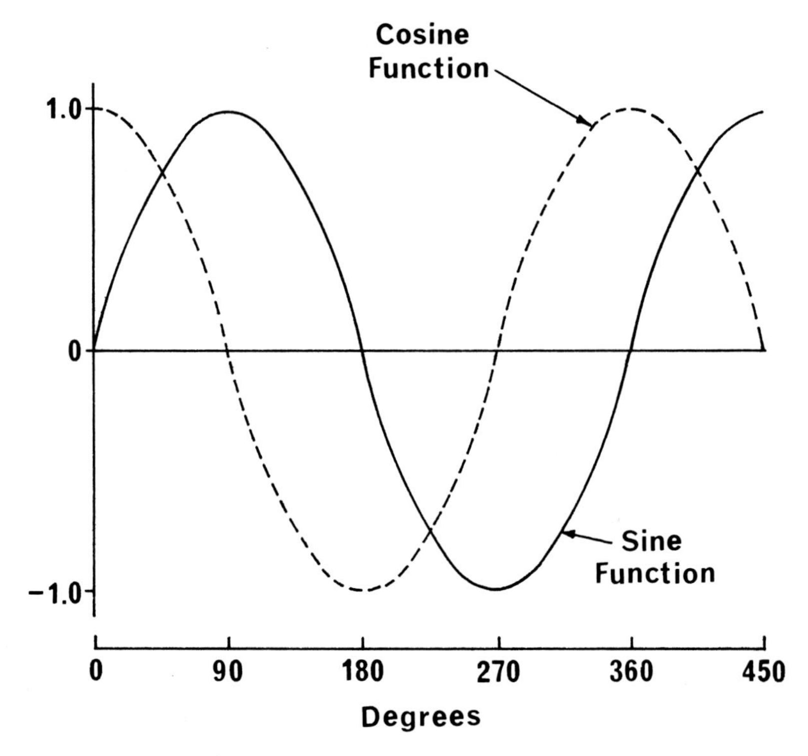

FIGURE A-2. *Graphic Representation of Sine and Cosine.*

Two trigonometric relationships which apply to any triangle also prove useful in biomechanical analyses. They are the sine law,

$$\frac{a}{\sin A} = \frac{b}{\sin B} = \frac{c}{\sin C}$$

and the cosine law,

$$a^2 = b^2 + c^2 - 2bc \cos A$$

in which a, b and c refer to the sides opposite angles A, B and C respectively.

When the digital computer is used for data processing or when linear and angular parameters are linked, angles must be expressed in radians rather than degrees. The magnitude of an angle (θ) in radians is calculated from the ratio of the arc length (s) to the radius (r).

$$\theta = \frac{s}{r} \text{ radians} \qquad \text{(see Figure A-3)}$$

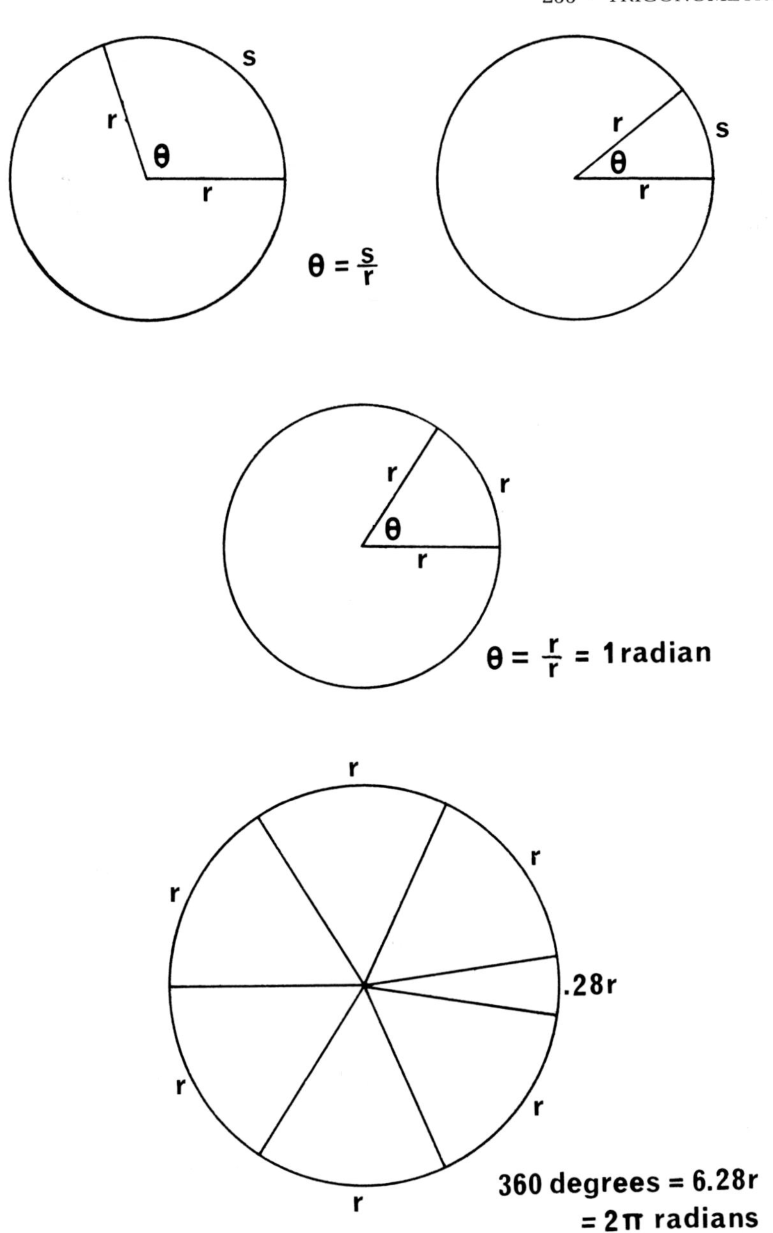

FIGURE A-3. *Relationship between Degrees and Radians.*

Table A-2. Trigonometric Reduction Formulae*

Degrees	*sin*	*cos*	*tan*	*cot*	*sec*	*csc*
$-\alpha$	$-\sin\alpha$	$+\cos\alpha$	$-\tan\alpha$	$-\cot\alpha$	$+\sec\alpha$	$-\csc\alpha$
$90° + \alpha$	$+\cos\alpha$	$-\sin\alpha$	$-\cot\alpha$	$-\tan\alpha$	$-\csc\alpha$	$+\sec\alpha$
$90° - \alpha$	$+\cos\alpha$	$+\sin\alpha$	$+\cot\alpha$	$+\tan\alpha$	$+\csc\alpha$	$+\sec\alpha$
$180° + \alpha$	$-\sin\alpha$	$-\cos\alpha$	$+\tan\alpha$	$+\cot\alpha$	$-\sec\alpha$	$-\csc\alpha$
$180° - \alpha$	$+\sin\alpha$	$-\cos\alpha$	$-\tan\alpha$	$-\cot\alpha$	$-\sec\alpha$	$+\csc\alpha$
$270° + \alpha$	$-\cos\alpha$	$+\sin\alpha$	$-\cot\alpha$	$-\tan\alpha$	$+\csc\alpha$	$-\sec\alpha$
$270° - \alpha$	$-\cos\alpha$	$-\sin\alpha$	$+\cot\alpha$	$+\tan\alpha$	$-\csc\alpha$	$-\sec\alpha$
$360° + \alpha$	$+\sin\alpha$	$+\cos\alpha$	$+\tan\alpha$	$+\cot\alpha$	$+\sec\alpha$	$+\csc\alpha$
$360° - \alpha$	$-\sin\alpha$	$+\cos\alpha$	$-\tan\alpha$	$-\cot\alpha$	$+\sec\alpha$	$-\csc\alpha$

*(Selby, S. M. (Ed): Standard Mathematical Tables. 19th Ed., Cleveland: Chemical Rubber Company, p. 188, 1971).

Since the units of measurement of r and s are the same (in., ft., cm., etc.), they can be cancelled out of the ratio, thus making radians a unitless quantity. When $s = r$, angle θ equals one radian. Therefore, one radian is defined as the angle subtended from the circle by an arc equal in length to the radius of the circle. Since the circumference of a circle is equal to $2\pi r$, there will be 2π radians in a circle.† Therefore,

$$2\pi \text{ radians} = 360 \text{ degrees}$$

$$1 \text{ radian} = \frac{360}{2\pi} = 57.295\ldots \text{ degrees.}$$

Other useful relationships include:

$$\begin{aligned} \pi \text{ radians} &= 180 \text{ degrees} \\ \pi/2 \text{ radians} &= 90 \text{ degrees} \\ \pi/3 \text{ radians} &= 60 \text{ degrees} \\ \pi/4 \text{ radians} &= 45 \text{ degrees} \\ \pi/6 \text{ radians} &= 30 \text{ degrees} \\ 1 \text{ degree} &= 0.017453\ldots \text{ radians.} \end{aligned}$$

SELECTED REFERENCES

Ayres, F.: *Theory and Problems of Plane and Spherical Trigonometry.* New York: Schaum, 1954.

Selby, S. M. (Ed.): *Standard Mathematical Tables.* 19th Ed., Cleveland: Chemical Rubber Company, 1971.

†π is the ratio of the circumference of a circle to its diameter which is constant for all circles, namely, 3.1415926 . . . . .

APPENDIX B

Elements of Vector Algebra

DETAILED ANALYSIS of noncoplanar biomechanical systems requires a basic understanding of the elements of vector algebra.

RIGHT HAND RULE

It is customary to use a right-handed coordinate system in most mechanical analyses (Figure B-1). If the fingers of the right hand are curled in the direction necessary to rotate the positive X axis toward the positive Y axis through the 90 degree angle between them, the thumb will point in the direction of the positive Z axis. Such a coordinate reference frame is termed right-handed.

UNIT VECTORS

Vectors, one unit in length, are utilized to indicate specific directions. The notation **i**, **j** and **k** indicates unit vectors directed along the positive X, Y and Z axes respectively (Figure B-2).

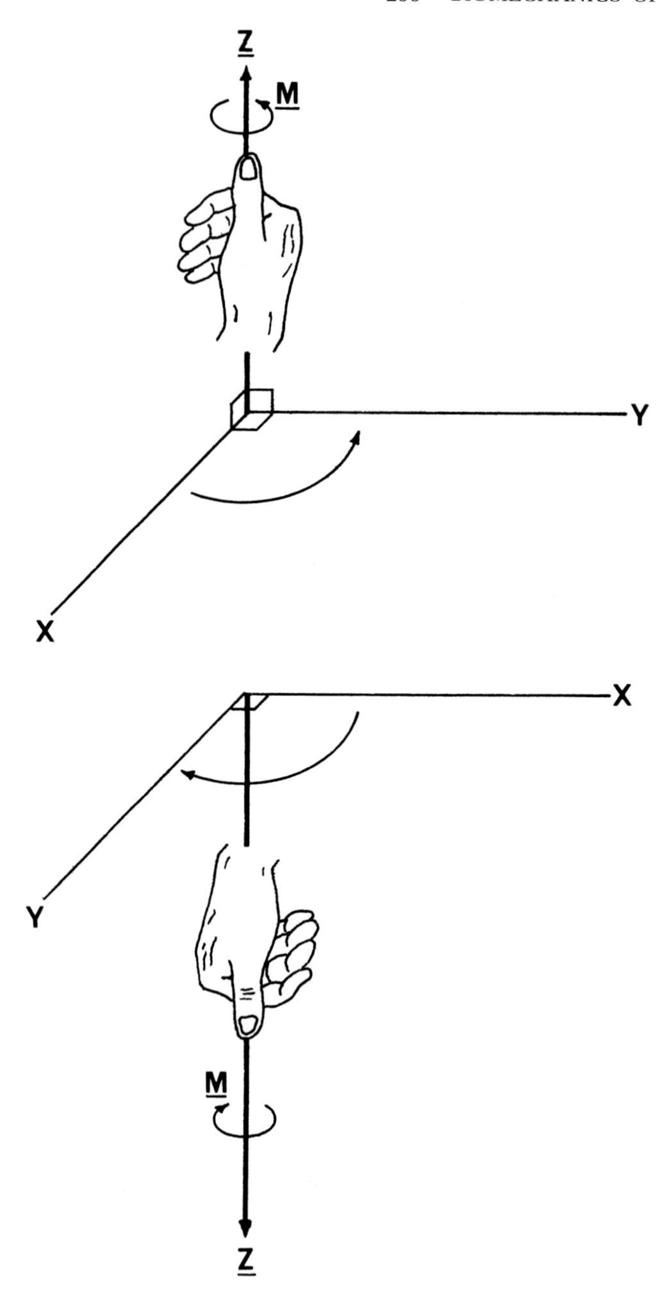

FIGURE B-1. *Right-handed Coordinate Reference Frames.*

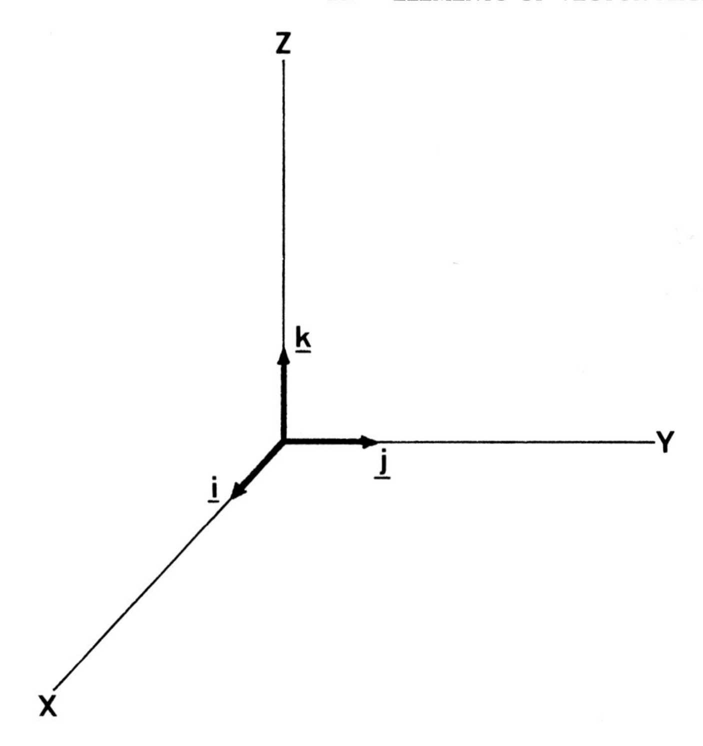

FIGURE B-2. *Unit Vectors.*

VECTOR NOTATION

A vector is often specified in terms of unit vectors **i**, **j** and **k** with each being multiplied by the appropriate magnitude. The vector notation for a line directed from $M(Mx,My,Mz)$ to $N(Nx,Ny,Nz)$ is:

$$\mathbf{MN} = (Nx - Mx)\mathbf{i} + (Ny - My)\mathbf{j} + (Nz - Mz)\mathbf{k}.$$

If the vector were directed from N to M, then it would be designated:

$$\mathbf{NM} = (Mx - Nx)\mathbf{i} + (My - Ny)\mathbf{j} + (Mz - Nz)\mathbf{k}.$$

Assuming that the coordinates of M and N were $(2,1,-2)$ and $(5,-3,10)$ respectively, the vector directed from M to N would be:

$$\begin{aligned}\mathbf{MN} &= (5 - 2)\mathbf{i} + (-3 - 1)\mathbf{j} + (10 + 2)\mathbf{k} \\ &= 3\mathbf{i} - 4\mathbf{j} + 12\mathbf{k}\end{aligned}$$

The magnitude of **MN** can be calculated using the three dimensional version of the Pythagorean theorem:

$$\mathrm{MN} = \sqrt{3^2 + 4^2 + 12^2} = \sqrt{169} = 13$$

MN can be expressed as a unit vector, **u**, by dividing each of its components by the magnitude of the vector. Thus,

$$\mathbf{u} = \frac{\mathbf{MN}}{MN} = \frac{3\mathbf{i} - 4\mathbf{j} + 12\mathbf{k}}{13}.$$

ADDITION AND SUBTRACTION

Vectors can be combined by the addition or subtraction of each of the three orthogonal components. For example, if

$$\mathbf{F} = Fx\mathbf{i} + Fy\mathbf{j} + Fz\mathbf{k} \quad \text{and}$$
$$\mathbf{G} = Gx\mathbf{i} + Gy\mathbf{j} + Gz\mathbf{k}$$

then $\mathbf{F} + \mathbf{G} = (Fx + Gx)\mathbf{i} + (Fy + Gy)\mathbf{j} + (Fz + Gz)\mathbf{k}$.

DOT OR SCALAR PRODUCT

The dot product of two vectors is, by definition, a scalar equal to the product of the magnitudes of the two vectors multiplied by the cosine of the smaller angle between them. Thus,

$$\mathbf{M} \cdot \mathbf{N} = MN \cos \theta \qquad \text{(Figure B-3).}$$

This gives the component of **M** directed along **N** multiplied by the magnitude of *N*. It is therefore a suitable vector operation for determining work which is equal to the component of force in the direction of the displacement multiplied by the displacement.

The dot products of unit vectors **i**, **j** and **k** are:

$$\mathbf{i} \cdot \mathbf{i} = (1)(1) \cos 0 = 1$$

Similarly, $\mathbf{j} \cdot \mathbf{j} = \mathbf{k} \cdot \mathbf{k} = 1.$

Then, $\mathbf{j} \cdot \mathbf{k} = (1)(1) \cos 90 = 0$ and likewise

$$\mathbf{i} \cdot \mathbf{j} = \mathbf{k} \cdot \mathbf{i} = \mathbf{j} \cdot \mathbf{i} = \mathbf{i} \cdot \mathbf{k} = \mathbf{k} \cdot \mathbf{j} = 0.$$

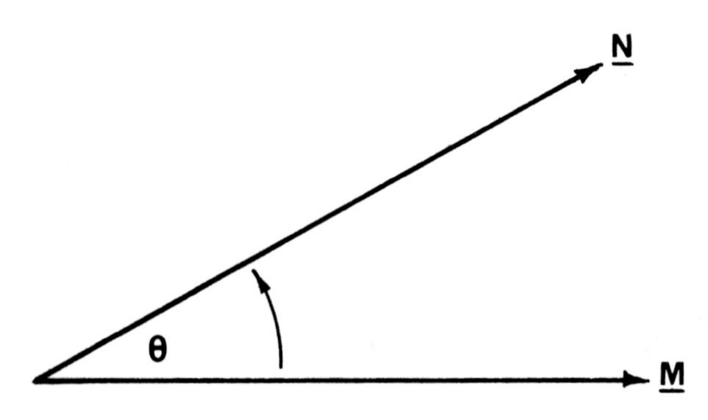

FIGURE B-3. *Dot Product.*

Using these identities to expand $\mathbf{M} \cdot \mathbf{N}$,

$$\begin{aligned}\mathbf{M} \cdot \mathbf{N} &= (Mx\mathbf{i} + My\mathbf{j} + Mz\mathbf{k}) \cdot (Nx\mathbf{i} + Ny\mathbf{j} + Nz\mathbf{k}) \\ &= MxNx + MyNy + MzNz.\end{aligned}$$

VECTOR OR CROSS PRODUCT

By definition, the cross product of two vectors **M** and **N** is a vector **V** acting in a direction perpendicular to the plane of the two original vectors in accordance with the right hand rule (Figure B-4). The magnitude of **V** is equal to the product of the magnitudes of **M** and **N** multiplied by the sine of the smaller angle between them. The cross products of the unit vectors **i**, **j** and **k** are:

$$\begin{array}{lll} \mathbf{i} \times \mathbf{i} = (1)(1)\sin 0 = 0 & \mathbf{j} \times \mathbf{j} = 0 & \mathbf{k} \times \mathbf{k} = 0 \\ \mathbf{i} \times \mathbf{j} = \mathbf{k} & \mathbf{j} \times \mathbf{k} = \mathbf{i} & \mathbf{k} \times \mathbf{i} = \mathbf{j} \\ \mathbf{j} \times \mathbf{i} = -\mathbf{k} & \mathbf{k} \times \mathbf{j} = -\mathbf{i} & \mathbf{i} \times \mathbf{k} = -\mathbf{j}. \end{array}$$

Using these identities, the cross product $\mathbf{M} \times \mathbf{N}$ can be expanded:

$$\begin{aligned}\mathbf{M} \times \mathbf{N} &= (Mx\mathbf{i} + My\mathbf{j} + Mz\mathbf{k}) \times (Nx\mathbf{i} + Ny\mathbf{j} + Nz\mathbf{k}) \\ &= MxNy\mathbf{k} - MxNz\mathbf{j} - MyNx\mathbf{k} + MyNz\mathbf{i} + MzNx\mathbf{j} \\ &\quad - MzNy\mathbf{i} \\ &= (MyNz - MzNy)\mathbf{i} + (MzNx - MxNz)\mathbf{j} \\ &\quad + (MxNy - MyNx)\mathbf{k}\end{aligned}$$

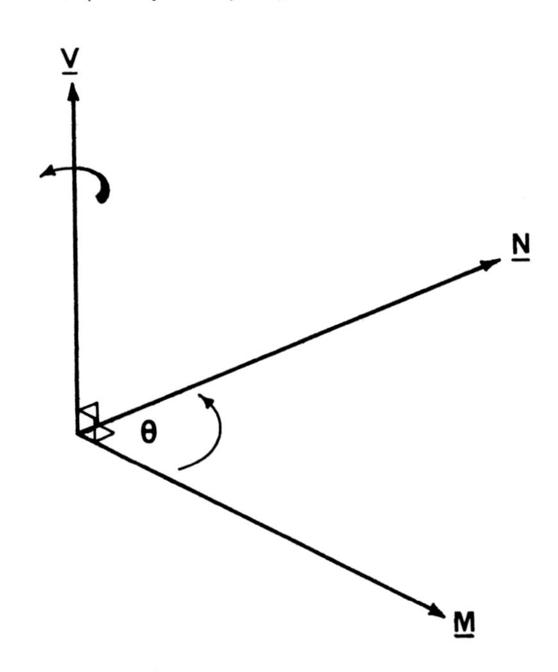

FIGURE B-4. *Cross Product.*

This relationship can be more easily expressed in the form of a three by three determinant.

$$\mathbf{M} \times \mathbf{N} = \begin{vmatrix} \mathbf{i} & \mathbf{j} & \mathbf{k} \\ Mx & My & Mz \\ Nx & Ny & Nz \end{vmatrix}$$

MOMENT OF A FORCE ABOUT A POINT

The vector product $\mathbf{r} \times \mathbf{F}$ represents the moment of force $\mathbf{F}$ about a point where $\mathbf{r}$ is any position vector directed from the point to the line of action of $\mathbf{F}$. Both magnitude and direction of the moment can be obtained by expanding the determinant:

$$\mathbf{r} \times \mathbf{F} = \begin{vmatrix} \mathbf{i} & \mathbf{j} & \mathbf{k} \\ r_x & r_y & r_z \\ F_x & F_y & F_z \end{vmatrix}$$

In the planar example shown in Figure B-5, the magnitude of the moment of $\mathbf{F}$ about point A is equal to the magnitude of the force multiplied by the perpendicular distance d. By trigonometry, it can be seen that

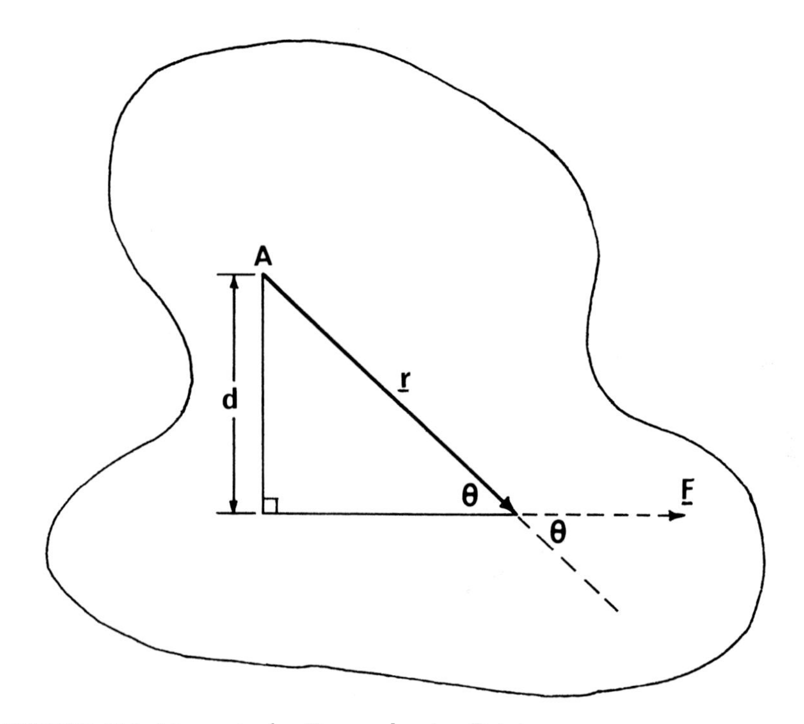

FIGURE B-5. *Moment of a Force about a Point.*

$$d = r \sin \theta.$$

It will be recalled that the magnitude of $\mathbf{r} \times \mathbf{F} = rF \sin \theta = Fr \sin \theta$.

MOMENT OF A FORCE ABOUT AN AXIS

The triple scalar product is utilized to determine the moment of force **F** about a particular axis **N**. The axis must first be expressed in unit vector form. Then, the moment of force **F** about axis **N** is $\mathbf{r} \times \mathbf{F} \cdot \mathbf{N}$ where **r** is any position vector joining **N** and **F**, directed toward **F**. In determinant form, the triple scalar product is:

$$\mathbf{r} \times \mathbf{F} \cdot \mathbf{N} = \begin{vmatrix} r_x & r_y & r_z \\ Fx & Fy & Fz \\ Nx & Ny & Nz \end{vmatrix}$$

The triple scalar product is the dot product of two vectors, one of which is the cross product of two other vectors. In computing the moment of force about an axis, the triple scalar product gives the component of rotation in the direction of the specified axis.

EXAMPLE

The following example illustrates some of the basic concepts of vector algebra which are required in biomechanical analysis.

Given: A 100-pound force passes through points $D(1,3,5)$ and $E(10,10,12)$ and is directed from D to E. The coordinates are specified in inches.

Find: The moment of the force with respect to an axis through $G(-1, -6, -12)$ and $H(-3, -10, -20)$. The axis is directed from G to H.

Solution:

(1) Express the force in vector form.

(a) Direction of the force

$$\begin{aligned} \mathbf{DE} &= (10 - 1)\mathbf{i} + (10 - 3)\mathbf{j} + (12 - 5)\mathbf{k} \\ &= 9\mathbf{i} + 7\mathbf{j} + 7\mathbf{k} \end{aligned}$$

(b) Express the direction as a unit vector.

$$\begin{aligned} \mathrm{U} &= \frac{9\mathbf{i} + 7\mathbf{j} + 7\mathbf{k}}{\sqrt{9^2 + 7^2 + 7^2}} \\ &= \frac{1}{\sqrt{179}}(9\mathbf{i} + 7\mathbf{j} + 7\mathbf{k}) \end{aligned}$$

(c) Multiply the unit vector by the magnitude of the force to determine the force vector

$$\mathbf{F} = \frac{100}{\sqrt{179}}(9\mathbf{i} + 7\mathbf{j} + 7\mathbf{k})$$

Thus the components of force in the three orthogonal directions are:

$$Fx = \frac{900}{\sqrt{179}} \qquad Fy = \frac{700}{\sqrt{179}} \qquad Fz = \frac{700}{\sqrt{179}}$$

(2) Calculate a position vector **r** joining the axis to the line of action of **F**

(a) Of the several possibilities available, namely, **GD, GE, HD** and **HE,** vector **GD** is arbitrarily chosen.

$$\begin{aligned}\mathbf{r} &= (1 + 1)\mathbf{i} + (3 + 6)\mathbf{j} + (5 + 12)\mathbf{k} \\ &= 2\mathbf{i} + 9\mathbf{j} + 17\mathbf{k}\end{aligned}$$

(3) Express the axis in unit vector form

(a) $\mathbf{GH} = (-3 + 1)\mathbf{i} + (-10 + 6)\mathbf{j} + (-20 + 12)\mathbf{k}$
$\quad = -2\mathbf{i} - 4\mathbf{j} - 8\mathbf{k}$

(b) As a unit vector, **N**

$$\begin{aligned}\mathbf{N} &= \frac{-2\mathbf{i} - 4\mathbf{j} - 8\mathbf{k}}{\sqrt{2^2 + 4^2 + 8^2}} \\ &= \frac{1}{\sqrt{84}}(-2\mathbf{i} - 4\mathbf{j} - 8\mathbf{k})\end{aligned}$$

(4) Use the triple scalar product to determine the moment of the force about the axis

$$\begin{aligned}M &= \mathbf{r} \times \mathbf{F} \cdot \mathbf{N} \\ &= \frac{1}{\sqrt{84}} * \frac{100}{\sqrt{179}} \begin{vmatrix} 2 & 9 & 17 \\ 9 & 7 & 7 \\ -2 & -4 & -8 \end{vmatrix} \\ &= \frac{100}{\sqrt{84}\sqrt{179}}(2(-56 + 28) - 9(-72 + 14) + 17(-36 + 14)) \\ &= \frac{100}{9.165 * 13.379}(-2 * 28 + 9 * 58 - 17 * 22) \\ &= \frac{9200}{122.62} = 75.03 \text{ pound-inches}\end{aligned}$$

SELECTED REFERENCES

Meriam, J. L.: *Dynamics*. New York: Wiley, 1966.

Shames, I. H.: *Engineering Mechanics—Statics and Dynamics*, 2nd Ed., Englewood Cliffs, N.J.: Prentice-Hall, 1967.

APPENDIX C

Data Smoothing in Biomechanics

IN BIOMECHANICS, the researcher is frequently concerned with the differentiation of experimental data. Displacement values obtained from film or electrogoniometers form the basis for the computation of velocity and acceleration. These important kinematic variables are the first and second time derivatives of displacement. Since each derivative represents the slope of the previous curve, irregularities in the original function are magnified out of proportion as subsequent derivatives are calculated.

It is impossible to eliminate all the sources of error from results obtained experimentally. Therefore, curve-fitting or data-smoothing techniques must be applied in an attempt to compensate for the inaccuracies and to approximate the true function.† Either manual or digital computer methods may be used for this purpose. The latter are the more widely accepted since there are a number of existing library programs to perform the necessary calculations quickly, objectively and accurately.

†For a detailed explanation of the theory of these methods, the reader is referred to books on numerical analysis.

MANUAL METHOD

The free-hand or manual method of smoothing data provides a crude approximation of the curve. The raw position values are first plotted against time. A smooth curve is then drawn passing through as many of the points as possible. A French curve may be used for this purpose. The next step is to construct tangents to the curve at equal intervals of time (Figure C-1). By definition, the tangent should just touch the curve at the given point and not intersect it. From appropriately scaled divisions on the graph paper, the slope of the line can be calculated.

$$\text{Slope} = \frac{\text{Change in Position}}{\text{Change in Time}} = \text{Velocity}$$

The velocity values are then plotted against time on a second graph and a smooth curve drawn to encompass as many of them as possible. To verify the accuracy of the results, the velocity-time function can be integrated graphically by determining the area beneath the curve and the integral compared with the corresponding displacement (Felkel, 1951). Similar steps can be followed to obtain acceleration data from the velocity-time curve. This graphical method,

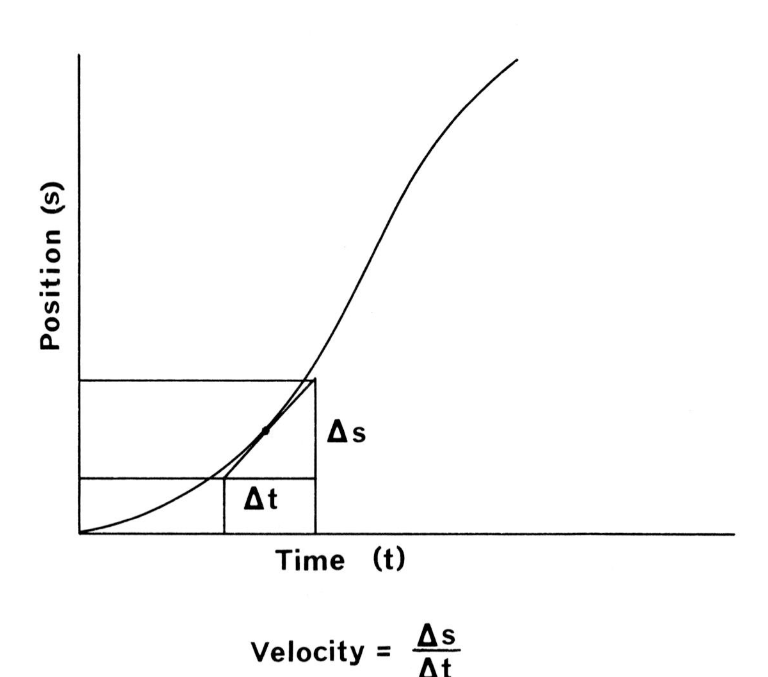

FIGURE C-1. *Graphical Differentiation.*

however, is rather time-consuming and requires well-trained personnel to achieve valid results. Even then experts frequently disagree on the best manual representation of the curve.

FINITE DIFFERENCES

Finite difference methods, which are based on Taylor series expansions, provide formulae for calculating first, second and higher derivatives of functions (James, Smith and Wolford, 1964). They are local approximation techniques since they utilize data points on both sides of the desired one. For example, the First Central Difference formula for calculating the first derivative of x_i as a function of time (velocity) is:

$$x_i' = \frac{x_{i+1} - x_{i-1}}{2(\Delta t)}$$

The second derivative of x_i with respect to time (acceleration) is:

$$x_i'' = \frac{x_{i+1} - 2x_i + x_{i-1}}{(\Delta t)^2}$$

in which x_i is a specific point;
x_{i+1} is the point to the right of x_i;
x_{i-1} is the point to the left of x_i; and
Δt is the time interval between two points.

With this method, it is necessary to employ First Forward Differences at the beginning since no data are available to the left of the first point.

$$x_i' = \frac{x_{i+1} - x_i}{(\Delta t)}$$

$$x_i'' = \frac{x_{i+2} - 2x_{i+1} + x_i}{(\Delta t)^2}$$

Similarly First Backward Difference equations must be used at the end.

$$x_i' = \frac{x_i - x_{i-1}}{(\Delta t)}$$

$$x_i'' = \frac{x_i - 2x_{i-1} + x_{i-2}}{(\Delta t)^2}$$

Second forward, central and backward difference formulae using two points on either side of x_i provide a more pronounced smoothing effect. Second Forward Difference Equations are:

$$x_i' = \frac{-x_{i+2} + 4x_{i+1} - 3x_i}{2(\Delta t)}$$

$$x_i'' = \frac{-x_{i+3} + 4x_{i+2} - 5x_{i+1} + 2x_i}{(\Delta t)^2}$$

Second Central Difference Equations are:

$$x_i' = \frac{-x_{i+2} + 8x_{i+1} - 8x_{i-1} + x_{i-2}}{12(\Delta t)}$$

$$x_i'' = \frac{-x_{i+2} + 16x_{i+1} - 30x_i + 16x_{i-1} - x_{i-2}}{12(\Delta t)^2}$$

Second Backward Difference Equations are:

$$x_i' = \frac{3x_i - 4x_{i-1} + x_{i-2}}{2(\Delta t)}$$

$$x_i'' = \frac{2x_i - 5x_{i-1} + 4x_{i-2} - x_{i-3}}{(\Delta t)^2}$$

Although several digital computer library programs are available for the calculation of derivatives employing the principles of finite differences, this method may not provide sufficient smoothing to offset adequately errors in data recording.

GRAPHO-NUMERICAL METHOD

The grapho-numerical method combines the best features of the manual and finite difference techniques (Felkel, 1951). The position data are initially smoothed manually. Position values are then read off this "best-fitting" curve and are used in the finite difference equations to calculate velocity and acceleration.

LEAST SQUARE APPROXIMATIONS

It is often desirable to express the relationship between variables as a mathematical equation. The raw data serve as the basis for deriving such an equation and the problem becomes one of selecting a suitable function. The method of least squares may provide an appropriate solution.

Consider a set of data points which represent position as a function of time. Although a number of polynomial equations of a given degree could be derived to approximate this relationship, there is one "best-fitting" curve in which the sum of the squared deviations is a minimum (Figure C-2). A first degree polynomial requiring two

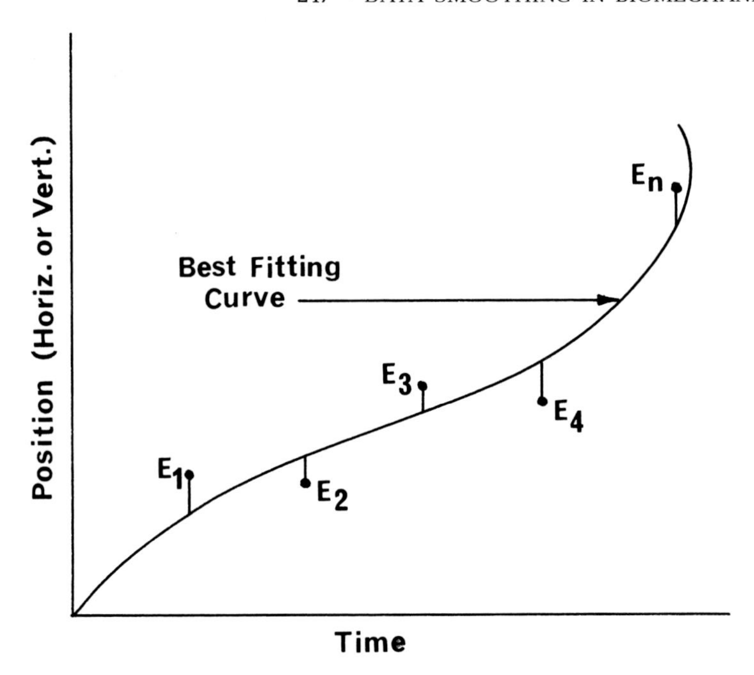

FIGURE C-2. *Least Square Curve Approximation.*

data points is a straight line; a second degree polynomial represents a parabolic or quadratic curve; and a third degree polynomial equation, a cubic curve. Given *n* data points, an *n-1* degree polynomial will provide a perfect fit passing through all *n* points. In most cases, however, the objective is to use as low a polynomial degree as possible. The least square error estimates will provide some guidance in choosing an appropriate degree to represent the data and introduce a certain amount of smoothing without obscuring actual trends in the data.

This method provides an objective means of curve fitting and does not require equally spaced data points. Computer programs for generating least square polynomial equations from raw data are readily available.† Subsequent differentiation of position-time polynomials to obtain velocity and acceleration is a straight-forward procedure.

SELECTED REFERENCES

Felkel, E. O.: Determination of Acceleration from Displacement-Time Data. Prosthetic Devices Research Project Report Series 11, Issue 16, Institute

†One such program is included in the Scientific Subroutine Package (SSP).

of Engineering Research, University of California, Berkeley, 1951 (September).

James, M. L., Smith, G. M., and Wolford, J. C.: *Analog and Digital Computer Methods in Engineering Analysis.* Scranton: International Textbook Company, 1964.

Spiegel, M. R.: *Theory and Problems of Statistics.* New York: McGraw-Hill, 1961.

Widule, C. J., and Gossard, D. C.: Data Modeling Techniques in Cinematographic Research. Res. Q. Amer. Assoc. Health Phys. Ed., *42*, 103–111, 1971.

Author Index

Subject Index

RETURNED
DATE
MAY 6 1975
RETURNED
24 1975
FEB 03 1995
AUG 24 2000
JUN 29 1976
DEC 19 1981
NOV 9 1982
DEC 04 2002
DEC 05 1983
MAY 08 1984
DEMCO 38-297